THE
TRIUMPH
OF
NUMBERS

ALSO BY I. BERNARD COHEN

Benjamin Franklin's Experiments (1941)

Roemer and the First Determination of the Velocity of Light (1942)

Science, Servant of Man (1948)

Some Early Tools of American Science (1950, 1967)

*Ethan Allen Hitchcock of Vermont: Soldier,
Humanitarian, and Scholar* (1951)

General Education in Science (with Fletcher G. Watson) (1952)

Benjamin Franklin: His Contribution to the American Tradition (1953)

Isaac Newton's Papers & Letters in Natural Philosophy (1958, 1978)

The Birth of a New Physics (1960, 1985)

A Treasury of Scientific Prose (with Howard Mumford Jones) (1963, 1977)

Introduction to Newton's "Principia" (1971)

Isaac Newton's Principia, *with Variant Readings*
(with Alexandre Koyré & Anne Whitman) (1972)

Isaac Newton's Theory of the Moon's Motion (1975)

Benjamin Franklin: Scientist and Statesman (1975)

The Newtonian Revolution (1980)

An Album of Science: From Leonardo to Lavoisier, 1450–1800 (1980)

Revolution in Science (1985)

Puritanism and the Ride of Modern Science: The Merton Thesis (1990)

*Interactions: Some Contacts between the Natural Sciences
and Social Sciences* (1994)

*Science and the Founding Fathers: Science in the Political
Thought of Thomas Jefferson, Benjamin Franklin, John Adams,
and James Madison* (1995)

Benjamin Franklin's Science (1996)

Newton: Texts, Backgrounds, Commentaries
(A Norton Critical Edition) (1996)

Howard Aiken: Portrait of a Computer Pioneer (2000)

THE

TRIUMPH

OF

NUMBERS

HOW COUNTING
SHAPED MODERN LIFE

I. BERNARD COHEN

 W. W. Norton & Company • New York • London

For information about permission to reproduce selections
from this book, write to Permissions,
W. W. Norton & Company, Inc., 500 Fifth Avenue, New York, NY 10110

Manufacturing by Quebecor World, Fairfield
Book design by Mary McDonnell
Production manager: Amanda Morrison

Library of Congress Cataloging-in-Publication Data

Cohen, I. Bernard, 1914–
The triumph of numbers : how counting shaped modern life / I. Bernard
Cohen.— 1st ed.
p. cm.
Includes bibliographical references and index.
ISBN 0-393-05769-0
1. Mathematical statistics—History. 2. Statistics—History. 3. Science—
History. 4. Science—Mathematical models—History. I. Title.
QA276.C53 2005
519.5'09—dc22

2004027322

W. W. Norton & Company, Inc., 500 Fifth Avenue, New York, N.Y. 10110
www.wwnorton.com

W. W. Norton & Company Ltd., Castle House,
75/76 Wells Street, London W1T 3QT

1 2 3 4 5 6 7 8 9 0

CONTENTS

Contents

Contents

THE
TRIUMPH
OF
NUMBERS

FOREWORD

The Triumph of Numbers, the late I. Bernard Cohen's last work, originally came with a long subtitle: "How Numbers Entered the Conduct of Life and of Government, the Understanding of Nature, and the Analysis of Society." Various editors, readers, and other commentators, myself included, thought that was too cumbersome, and the less baroque subtitle that the book now carries—"How Counting Shaped Modern Life"—was put in its place.

I suspect Bernard would have objected to these changes in terminology. The verb in his subtitle was "entered," not "shaped"; and the cases he chose to illustrate the making of the numerical worlds of government, nature, and society highlight differences among those domains that the all-encompassing phrase "modern life" obscures.

That said, the revised subtitle catches the thrust of his work better than his own did in two important respects. First, the history he wished to write is a phenomenon he clearly saw—and clearly wanted his readers to see—as a fact of modern life in general. "We live in a world of numbers," he says at the very start of Chapter 1, a veritable "sea" that is of recent origin and, like the air we breathe, now exists "everywhere." Second, *The Triumph*

11

of Numbers is a book with a broad thesis (also announced at the beginning of Chapter 1, but with less fanfare) about the role of something like techniques of counting in the creation of this modern ocean; that thesis holds that numbers would not be all around us today were we not also awash in a stream of inferences, predictions, concepts, theories, calculations, and findings of fact based on them.

The stream in question is, of course, not the only current flowing into our numerical sea. It does, however, stand out from the other major tributaries that Bernard lists—the long-term growth of cities and commerce, the mounting demands of wars on states and societies, and the seemingly relentless expansion of government bureaucracies—in that it appears to have no pre-modern antecedents. "Through all of recorded history, every organized society or system of government has relied on numbers in some way," he tells us, but "no systematic analyses of these numbers occurred until well into the seventeenth century, the age of the Scientific Revolution." Even then they were more the exception than the rule and, all other things being equal, might have remained so.

Why they did not is the puzzle around which *The Triumph of Numbers* is organized, or around which it comes to be organized after a brief excursus on the use of numbers as keys to historical problems. Apart from that detour, the discussion proceeds fairly well apace; having identified a major intellectual watershed and located us on the near side of it, Bernard offers a set of accounts of how we managed to consolidate our position there, that is, how numbers of all sorts came to seem self-evidently worthy—and in need—of systematic analysis.

The story turns out to be a complicated one, full of odd byways and more than a few dead ends, because numbers have lent themselves to a variety of analytic aims and purposes. As a longtime student of early modern physics and astronomy,

Bernard predictably gives initial pride of place to the seventeenth-century discovery that numbers could be used to state general laws of nature and frame questions to test scientific theories. But his essay has mainly to do with the rise (and sometimes fall) of other interpretive programs, such as schemes for arriving at "a reasonable estimate of a significant social number"—say, the size of London's population or the wealth of France—from compilations of "specific numerical data"; for translating "commonsense notions about morality into mathematical language"; for constructing probabilistic evaluations of voting procedures; for arming medicine with mathematical measures of therapeutic efficacy; for generating evidence of regularities in human behavior from social and vital statistics; for displaying statistical data in ways that made a case for hospital and sanitary reform. On the other side of the coin less obviously rewarding numbers were used to link Martin Luther's name to the number of the beast, to produce magic squares, and to work out the rule according to which lilacs bloom.

Historians of science will recognize many of Bernard's stories and most of their protagonists. As his bibliography shows, there is now a substantial scholarly literature on such topics as Kepler's astronomy and Galileo's mechanics, political arithmetic from John Graunt and William Petty on, Laplace's probability theory and its applications, the numerical method of Pinel and Louis, Quetelet's social physics (although not his horticultural ideas), and the statistical enthusiasms of Florence Nightingale and William Farr, not to mention numerology and number mysticism on the one hand and the anti-quantitative polemics of Charles Dickens (*Hard Times*) and Thomas Carlyle ("Chartism") on the other.

The effect, however, of having them brought together in one place is striking. It helps that, while concerned throughout to depict them as parts of a larger whole, Bernard makes much of the contrasting, even contradictory, visions they represent. He does so by way of elaborating on and sharpening his point about the numerical worlds of nature, government, and society being built on very different principles. Simply put, the message is that it was not mathematical physics and astronomy, or political arithmetic, or social statistics that triumphed, but numbers. Nor did any one particular view of how they were to be used win out, and certainly not the grand seventeenth-century hope that, when properly analyzed and interpreted, numbers would reveal universal laws.

The Triumph of Numbers essentially ends with a downbeat comment on the failure of "sociological analogs of physical laws" to materialize, despite repeated assertions that they were just waiting to be found. That judgment suggests a final observation that Bernard would have appreciated, but which I doubt ever occurred to him. It is that his study resembles Albert Hirschman's well-known essay *The Passions and the Interests*, or at least might have done with something like a suitably modified version of its subtitle, *Political Arguments for Capitalism before Its Triumph*.*

For this book is ultimately a canvas of arguments, although not always political, for and against numbers before their triumph, indeed before most of the numbers that triumphed really existed, even before the machinery required to produce them was in place.** Often in this book Charles Dickens comes across as something of a numerical Luddite. But, perhaps he should have

---

*Albert O. Hirschman, *The Passions and the Interests: Political Arguments for Capitalism before Its Triumph* (Princeton, New Jersey: Princeton University Press, 1977).

**For an especially trenchant account of how much remained to be done on the latter score at the end of the nineteenth century, when *The Triumph of Numbers* ends, see the Introduction to J. Adam Tooze's excellent monograph, *Statistics and the German State, 1900–1945* (Cambridge, UK: Cambridge University Press, 2001).

the final word, for one of his more famous novels could have provided Bernard with a fitting subtitle—not *Hard Times* but *Great Expectations*.

Peter Buck
Harvard University
August 2004

I   A WORLD OF NUMBERS

NUMBERS EVERYWHERE

We live in a world of numbers. Some of these numbers lie hidden from us in the operations of government, the conduct of business and finance, the activities of science and engineering, and even in some aspects of daily life. Other numbers command our attention in the fluctuating price of gasoline and our monthly bank balances and credit-card statements. We encounter numbers in newspaper headlines and TV newscasts—CLI, GNP, the Dow (or Dow Jones) and Nasdaq. Sports fans discuss "stats" such as RBI; headlines declare the size of the national debt or the annual federal budget. A mysterious numerical quantity called APR appears on our monthly credit-card statements. And each day we face omnipresent numbers in the costs of food, clothing, rent, medical care, insurance, and college tuition.

Another all too obvious way in which we become aware of numbers lies in the difference between our earned wages or salary and the actual "take home" pay. Each week or each month our paycheck is docked for such items as social security, withholding for federal and state income taxes, pension, health insurance, and possibly union dues. Such a roster of numerical items makes us long for "the good old days" before the Sixteenth Amendment introduced the personal income tax.

Perhaps most poignant, the experience of the presidential election of the year 2000, with the result hinging on the final numbers from the Florida count, brought home to every citizen the importance of numbers in our political system. That same year the decennial census was taken—that statistical information yielding projections of cultural and economic trends in our society, among other things.

When did this invasion of numbers begin? History records the ever-increasing need for ordinary men and women to be able to count and to do simple arithmetic. With the introduction of our system of Hindu-Arabic numerals calculations became easier to perform than with the cumbersome Roman numerals. And indeed, the great increase in literacy that occurred after the invention of the printing press was accompanied by a similar increase in the ability to do simple sums in arithmetic. In early agricultural communities, a farmer or laborer did not need to be skilled in arithmetic because the general practice was to exchange services for goods or to trade one type of good or service for another. In cities, however, conditions of life were very different. A workman would be paid, at least in part, in monetary wages. With increased urbanization, there grew a need for city-dwellers to budget their earnings, to allot definite sums of money for rent, for food, for clothing. With the rise of industrialism and the growth of cities, conditions of life—even for simple people— became more and more numerical. Double-entry bookkeeping was devised and introduced in commercial transactions. Governments also became more numerically dependent as taxes of various sorts were introduced to pay for the expenses of war and the costs of maintaining an ever-increasing bureaucracy.

Although, over several centuries, enumerators collected numbers relating to such social phenomena as births, deaths, and marriages, no systematic analysis of these numbers occurred until well into the seventeenth century, the age of the Scientific Revolution.

NUMBERS IN HISTORY

Through all of recorded history, every organized society or system of government has relied on numbers in some way. A state or city has certain obvious numerical needs. Number problems arise in levying taxes, paying for armies, and other national or group activities. Indeed, historians and archaeologists have found such records going back to remotest antiquity. We know that several thousand years ago, some two thousand years before the beginning of the Christian Era, Babylonian scribes were busy with tax records kept on clay tablets. Their work survives in thousands of these tablets containing information about taxation and other business transactions. Some sense of the way the ancients dealt with numerical quantities can be gained from a royal mace of about 3500 B.C.E. This object, dug up in Egypt and now on display at a museum in Oxford University, records the capture of 120,000 prisoners together with 400,000 captive oxen and 1,422,000 goats.[1] Perhaps the numbers are exaggerated, but we can, even so, learn from this object that the ancient Egyptians were able to write very large numbers.

The history of society and civilization during the two millennia of the Christian Era shows how important numerical considerations have been in religious affairs, in the conduct of the state and the organization of society, in realms of commerce, and in architecture. For instance, surveyors used numbers to compute areas, a continuing practical art at least since the days of Egyptian civilization. Ancient Muslim mosques and early Christian churches give permanent testimony to skill in using numbers in the determination of the four cardinal points. And, of course, construction of the pyramids in Egypt entailed extreme numerical acumen.

In fact, because the pyramids are so familiar to all of us, we will study their construction as a case in point. We begin by considering the numbers associated with the Great Pyramid of Gizeh in ancient Egypt, also known as the pyramid of Cheops or Khufu. This pyramid was constructed around 2500 B.C.E., some 5,000 years ago. Its size is enormous. The base is a square that covers a little more than 13 acres, an area large enough to contain side-by-side St. Peter's of Rome and the cathedrals of Milan and Florence, together with Westminster Abbey and St. Paul's Cathedral of London.[2] The height is almost 500 feet, about the same as a building 40 stories high. Until the construction of the Eiffel Tower at the end of the nineteenth century, this pyramid was the tallest building in the world, a mammoth structure built of 2.3 million stones, each weighing from 2 to 15 tons each. The average weight of the stones is about 2.5 tons, the same as the weight of a large family van or an old-time Cadillac limousine.

In considering the subject of numbers and the pyramids, a distinction must be made between two kinds of pyramid numbers. For several centuries at least, there has existed a cadre of individuals who have seen in the dimensions and orientations of the pyramids a key to some sort of mystical knowledge. Their activity is sometimes referred to as pyramidiocy and the practitioners of this numerical game are known as pyramidiots.[3] The number includes some distinguished scientists, notably Sir Charles Piazzi Smyth, a nineteenth-century Astronomer Royal of Scotland.

Here we are not interested in how the dimensions or orientations of the pyramids may provide a key to secret knowledge. Rather, the goal is to show how considerations of numbers raise basic questions concerning technology several thousand years ago. We shall see that the use of numbers raises questions about the actual stages of construction of these extraordinary buildings.

During the Napoleonic campaign in Egypt in 1798 a famous battle was fought outside Cairo in Embaba, in view of the pyramids, and is known as the Battle of the Pyramids. A medal commemorating the French victory carries the words that Napoleon is said to have used while exhorting his troops: "Soldats! Du haute de ces pyramides quarante siècles nous contemplent"—"Soldiers! From the height of these pyramids forty centuries are watching us."[4] Napoleon, skilled in mathematics, had an uncanny sense of numbers and a deep sense of history. At that time, the date of construction of the pyramids had not yet been determined, but Napoleon hit it right on the nail. His "forty centuries" or 4,000 years would set this date at around 2200 B.C.E. Today we know that the date of the Great Pyramid was about 2500 B.C.E., just some 300 years earlier than Napoleon's estimate.

It is reported that, after the battle, Napoleon and his staff officers visited the Great Pyramid. While the more adventurous officers climbed to the top, Napoleon himself was content to rest in the shade of the pyramid at its base, toying with numbers. When the officers descended and rejoined him, Napoleon announced that he had made a calculation of the amount of stone in the pyramid. There was enough, he said, to build a stone wall 3 meters high and 0.3 meter thick (about 10 feet high and 1 foot thick) that would enclose the whole of France. One in the French group was Gaspard Monge, a great mathematician and the inventor of projective geometry. He made his own calculation and is said to have declared that Napoleon had been quite right in his estimate.[5]

The volume of a pyramid is equal to one-third of the product of its base times its altitude. The base of the Great Pyramid is a square whose sides measure 229 meters (756 feet), thus covering an area of 5.37 hectares (13.1 acres) or 52,441 square meters. The height is 146 meters. Hence the volume is $1/3 \times (52,441 \times 146) =$ 2,552,000 cubic meters. Assuming that the pyramid is solid—that is, that there are no significant hollow spaces or chambers—then it is easy to compute the length of a wall that could be made out of this mass of stone and that would be 3 meters high and 30 centimeters (0.3 meter) thick, as specified by Napoleon.

If all the stone in the pyramid were cut up and made into a wall that was 3 meters high and 0.3 meter thick, it would produce a wall 2,836,000 meters or 2,836 kilometers long. The reason is that 2,836,000 × 3 × 0.30 = 2,552,000 cubic meters, the total volume of the pyramid.

France has roughly the shape of a rectangle, almost a square, a little bit longer from north to south (approximately 770 kilometers, as measured from Montpelier to Rheims) than from east to west (approximately 700 kilometers, as measured from Nantes to Besançon). Hence the perimeter is approximately (2 × 770) + (2 × 700) kilometers or some 2,940 kilometers.

These numbers show us that Napoleon's estimate was of the right order of magnitude. The pyramid contains enough stone to compose a wall 2,836 kilometers long and the perimeter of a rectangle large enough to hold all of France is some 2,940 kilometers, an agreement within about three percent.

Of course, if the wall in question did not compose a rectangle large enough to hold all of France, but had to follow the ins and outs of France's borders and coastline, the wall would have to be much longer. But I doubt if even a Napoleon could make an accurate estimate of the length of such a snakelike curved perimeter.

The Great Pyramid's stones were cut to fit into place with near-jeweler's precision. They were quarried on the opposite side of the Nile, many miles away, and had to be transported across the river and carried to the building site, where they were raised into position and fitted into the pyramid structure with exactness. It is estimated that some 100,000 laborers were employed in this construction and that it took them 20 years or more to complete the task. Some ancient authorities say that it took 10 years just to build a ramp so that stones could be brought to positions as high as 40 stories in the air and that another 20 years were needed for the construction of the pyramid itself.

What does this recital of numbers tell us? First of all, that the Egyptians must have had a highly skilled accounting or book-keeping system, first to plan the preparation of this large quantity of stones cut at the quarry, then to ensure the orderly transport of these stones from the quarry to the building site, and finally to solve the engineering problem of putting each huge stone into its proper place in the eventual structure.

There was also the need to plan for food, drink, and shelter for the laborers. And of course the perfect fit of the stones, the perfection of the architectural design, and the long life of this pyramid and others give mute testimony to the architectural skill of the designers and the high level of craftsmanship of the artisans. Which of our structures will similarly survive for 4,000 years!

In recent years the problem of pyramid construction has been tackled in a new way by Dr. Richard H. G. Parry of Cambridge University. Dr. Parry is not an Egyptologist but a member of the Soil Mechanics Group in the Department of Engineering. As an engineer interested in pyramid construction, Dr. Parry dealt with two fundamental questions: First, how did the Egyptians transport these huge stones from the quarry to the building site, and second, how did they move these stones up a ramp to their position in the eventual structure? He was concerned with the time it would take to get the stones into place.

The Great Pyramid was completed in about 20 years. Therefore, since 2,300,000* stones were set in place in 20 years, the number of stones must, on average, have been 2,300,000/20, or

---

*These numbers give a correct order of magnitude but there are two reasons why we cannot say exactly how many stones are present in the pyramid. One is that there are some empty spaces in the interior burial chambers, chapels, etc. The second is that some of the interior space would have been filled with rubble.

115,000 stones per year. The laborers presumably worked every day without a day off, 365 days a year. Thus, the average number of stones put in place in a single day must have been at least 115,000/365, or 315 stones. Assuming that the laborers would have worked on average some 10 hours each day, we can calculate that the number of stones set in each hour must have been 315/10, or about 30. In other words, simple arithmetic applied to the numbers shows that the building of the pyramid must have involved setting 30 stones in place, on average, in every hour of the working day. This conclusion can be stated simply as follows: a giant stone was put into the structure every two minutes. How could this have been done?

There must have been enormous ramps along which the stones could be hauled from ground level to the place where they were fitted into the structure. Yet this problem is more complex than simple transport—the ancient Egyptians had no wheeled vehicles and so could not have used carts to haul the stones into position. For many years, it was supposed that the stones were hauled by laborers up ramps along a track lubricated by mud from the riverbed or along log rollers continually inserted at the front of sleds or sledges or skids—possibly using crowbars.

As a practicing engineer, Dr. Parry was impressed by the numbers and by the scale of the problem. He saw that the sledge method of transport could not account for a delivery of stone blocks, some weighing as much as 15 tons, at the rate of one every two minutes. Dr. Parry eventually produced a wholly new solution to the problem. In his reconstruction, the same mode of transportation is used to get the stones from the quarry to the building site and to move the stones up the ramp. Dr. Parry's solution to the pyramid problem was inspired by observing some models of the tools used in a temple from the New Kingdom, now in the Metropolitan Museum of Art in New York City (see figure 1.1). These wooden cradles could be fitted to each end of a

FIGURE 1.1 Wooden cradle used to transport building stones, from an Egyptian temple of the New Kingdom. *Courtesy of the Metropolitan Museum of Art, Gift of the Egypt Exploration Fund, 1896 (96.4.9)*

stone, creating a huge roller. These rollers could easily be moved along a level grade, as from the quarry to the pyramid, and readily rolled up the ramp. Dr. Parry constructed models of these rollers and showed that they gave an effective solution for the transport problem.[6]

Some supporting evidence for the use of wooden cradles or baskets is found in a statement of the ancient Greek historian Herodotus, writing about 500 B.C.E., some 2,200 years after the Great Pyramid was built. According to Herodotus, the pyramid builders of ancient Egypt made use of "contrivances composed of short pieces of timber for moving stones upwards." A small-scale model of a stone encased in circular runners is shown in figure 1.2.

Parry's solution was not entirely theoretical. A full-scale test of his circular runners was carried out near Tokyo by the Obayashi Corporation (see figure 1.3), using concrete blocks 0.8 meter square and 1.6 meters long, each weighing 2.5 tons. In these tests, laborers tried rolling a block along level roads and up

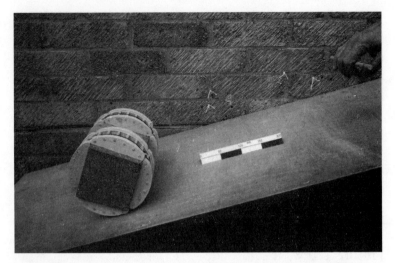

FIGURE 1.2 A model of a stone fitted with cradles. *Courtesy of R. G. H. Parry*

ramps with slopes of one-in-ten and one-in-four. The report of the trials showed that it took only three men to move the blocks encased in rollers along a level path, whereas 20 to 30 men were needed to move such a block on a sledge. Furthermore, 16 to 20 men were able to move a 2.5-ton stone up a one-in-four gradient (approximately that of the pyramid) when encased in rollers, whereas 60 to 80 men were needed with a sledge.

I asked Dr. Parry how the Obayashi Corporation happened to make an experimental test of his ideas. He told me that in 1995 he had been lecturing in Cairo (to a group of engineers) on the construction of the pyramids. His talk was based on data from tests made with small models. A Japanese friend in the audience then decided to make a full-scale field test. He was able to persuade the Obayashi Corporation, "one of the 'Big Five' construction companies in Japan," to make a trial to see whether Dr. Parry's proposed methods of moving blocks of stone would actually work.

FIGURE 1.3 A test of Parry's cradle-transport hypothesis, carried out near Tokyo by the Obayashi Corporation. *Courtesy of R. G. H. Parry*

The final test was one of time. Could a team of 20 men haul a 2.5-ton stone up a slope about 50 feet long in only a few minutes as the numbers obviously require? In the Japanese trials it was found that Parry's method of hauling up a slope enabled a team of 20 men to pull such a stone up a 50-foot slope in just one minute. Parry was pleased by the results. "I was pretty confident," he is reported to have said, "but you can never be 100 per cent confident about these things." He believed he had at last explained how the Egyptian builders "managed to handle very large blocks and place them in a very short period of time."

Dr. Parry's solution of the mode of transport of the stones in the pyramids fits the time constraint of getting one stone in place, on average, every two minutes. This mode of transport also solves the logistical problems of construction. In the present context, what is most significant about Dr. Parry's work is that we see how numbers provide a key to our understanding of ancient technologies and bygone civilizations.

NUMBERS IN THE BIBLE:
THE SIN OF KING DAVID

Readers familiar with the Bible know that the Old Testament contains many references to numbers. Indeed, one of the "books" is actually entitled the Book of Numbers. Some of the numbers in the Bible provide puzzles for interpreters. One such is the age of Methuselah, given as 969 years. Then there is a puzzling reference to the great sin of King David.

The numbers in the Bible have been used to compute the age of the earth, the source of major contention in today's debates between creationists and evolutionists. One very influential passage dealing with numbers occurs in the apocryphal Old Testament book the Wisdom of Solomon. Chapter 11, verse 20, records that God "created all things by number, weight, and measure." This early statement has reappeared in the twentieth century in a slightly different version: that "God is a mathematician." The Bible has also been a major sourcebook for numerologists. These number jugglers sought for meanings by converting various names into numbers and then showing the relation of these numbers to numbers found in the apocalyptic books of the Bible.

Yet the Bible records many head counts or censuses beyond the two found in the Book of Numbers, and most were made without any mystical intent. For some of them, the primary goal was to determine the number of males of fighting age. The first of these is recorded in Exod. 30:11–14, after the Israelites had been freed from bondage in Egypt. The Lord ordered this count in a command to Moses. The total was 603,550 males aged "twenty years and upwards." As with other Biblical censuses, taxes were collected as part of the enumerating. In this census, as in the case of the famous census recorded in Luke 2:1–8, the people being counted went to the enumerators, rather than having enumerators go out to count the people.

A year after the first census (Num. 1:2–19), the Lord once again commanded Moses to make a count to determine the number of males who were "twenty years old and upward," that is, all who were "able to go forth to war." For this census, the Bible records not only a gross total, as in the previous census, but the number. The total is exactly the same as the year before, 603,550—a fact that may indicate an estimate and not an actually counted number.

Another enumeration associated with a tax collection, is—because of the associated event—perhaps the most frequently cited census ever undertaken. The details are recorded in the New Testament where we are told that the Emperor Augustus issued a decree "for a registration to be made throughout the Roman world." This is presented as "the first registration of its kind." It was required that "everyone make his way to his own town." Accordingly, Joseph, accompanied by his wife Mary, departed from Nazareth where they were living and went to "the city of David, called Bethlehem, because he was of the house of David by descent."

Less well known to readers today is another census recorded in the Bible, which as a historical force may well be one of the most significant enumerations ever made. The details are recorded twice in the Old Testament, first in 2 Sam. 24:1–25 and then in 1 Chron. 21:1–30. According to the account in Samuel, King David was "incited" to give orders that "Israel and Judah should be counted." For reasons that are obscure, the taking of this census came to be viewed as evil in some way. In Samuel, the inciter is identified as "the anger of God." In Chronicles it is "Satan" who "stood up against Israel and provoked David to number Israel." In both accounts, David called Joab, the commander of the army, and gave him instructions to "go number Israel" and "bring the number" to David so that he might "know it." Joab evidently knew that this was an evil project and, according to Samuel, was reluctant to undertake this assignment. In Chronicles, Joab additionally asks David why he "will be a cause of trespass."

The enumerators were "officers of the army." They completed the count in nine months and 20 days, whereupon (according to Samuel) they "reported to the king the total number of people." In Israel there were 800,000 "able bodied men capable of bearing arms" and 500,000 in Judah. These would seem to be round numbers, estimates of the "valiant men that drew the sword," not an exact—or complete—enumeration. One reason for assuming that this is an estimate rather than a count is suggested in Chronicles: although Joab began the enumeration, he did not finish the job. The reason given (1 Chron. 27:24) is Joab's fear "of the wrath for it against Israel." The total given in 1 Chronicles is 1,570,000, a rather different number from that given in Samuel.

In the Chronicles account, "David's conscience smote him" (1 Chron. 21:8). And so "he said to the Lord, 'I have done a very wicked thing,' " and prayed to the Lord to "remove thy servant's guilt." The reply listed three possible punishments whereby he could atone for this misdeed; David was to choose one. These punishments were so dreadful, of such terrible magnitude, that they give us a measure of the enormity of David's sin in having ordered that a census be made. These punishments were: three years of famine, three months of defeat and slaughter by enemies, or three days of pestilence. David chose the last of these and the result was the death by plague of 70,000 people.

This story is disturbing in many ways. First, 70,000 innocent people had the bitter fate to die in penance for the sin of their king in making a count of the population. Then the question arises as to why David sinned by ordering a census. The Bible does not record that David or anyone else had been given a specific injunction against making such a count. The fact that, in

Chronicles, Satan incites David indicates the sinful nature of the enumeration, but does not tell us why this was so. This sinful nature of a census is stressed in both accounts by Joab's initial hesitation: he was aware of the sin in making such a census.

Our reading is further troubled by the fact that in 2 Chron. 2:17 we are told that Solomon, David's son, "took a census of all the aliens resident in Israel." This count was "similar to the census which David his father had taken." There is no hint here, a generation later, that taking this census might be a sinful act or that Solomon had so transgressed that 70,000 people should die as punishment!

A little later (Ezra 2:1–34), a count is recorded that was made of exiles returning from Babylonian captivity. Later still (Neh. 7), there is a detailed count of the population, again with no such consequences as followed upon David's census.

It is generally thought—by the scholars annotating the *New Oxford Annotated Bible*, for instance—that David's "sin" consisted of some human transgression against God's exclusive sovereignty over Israel. But this is presumption: no concrete rationale, even one along these lines, is ever made explicit. Nor is it explained why an enumeration should have been sinful when made by David, but not when it was made by Solomon or other leaders. The theological rationale of David's sin remains murky.

CONSEQUENCES OF DAVID'S SIN

David's census has been important historically because it implied that making a census could be considered a sin. In the seventeenth and early eighteenth centuries, people still knew their Bible and tended to have a literal belief in what they read. The story of David's census and the subsequent punishment by plague had important repercussions in these centuries, becoming a factor that prevented the taking of an accurate census in Britain and in her colonies. In England, at that time, many people were aware of the importance of having a population count. The evident opposition to having such a census was the fear of incurring the consequences of the sin of David. Cobbett's *Parliamentary History of England* records the opposition, as late as 1753, to a bill being considered that would require a complete enumeration of the British population. One member of Parliament cited the many letters he had received from constituents opposing the bill for fear of the kind of punishment meted out to David.[7]

In America in the Colonial period, a marked resistance to an accurate census stemmed in part from the fact that such data could provide a base for further taxation. Colonial administrators therefore found ways to avoid carrying out instructions for counting the population. They evidently reckoned the size of the population from existing tax lists and called them "censuses." Yet a study of this question has shown that opposition to a census also arose in part because the colonists were religious people who believed the stories of the Bible. Accordingly, they would have had a fear that a census would displease the Almighty and cause them to be punished, just as the people of Israel had been punished for the sin of David.[8] Two examples will serve as illustrations.

In 1634 Governor John Winthrop of the Massachusetts Bay Colony estimated rather than counted the population and referenced David's sin. In a letter of 22 May 1634, Winthrop wrote to Sir Nathaniel Rich that "for the number of our people, we never took any surveigh of them, nor doe we intend it, except enforced through urgent occasion." The reason was that "David's example stickes somewhat with us." He therefore only gave an estimate, "I esteeme them to be in all about 4000 Soules and upwards."[9]

In 1712, in a letter to the Lord of Trade, the governor of New York blamed the imperfections of the census of 1712 on the fear of God's wrath and, in a report, claimed that an earlier count had been followed by excessive sickness in the colony. In 1726, in a report to the British Board of Trade, the governor of New Jersey referred to King David's sin in relation to a census taken in New York three years earlier. He said that he would have ordered a census in New Jersey, but he had been advised that this would "make the people uneasy," that they "would take it for a repetition of the same sin that David committed in numbering the people, and might bring on the same judgments."[10]

The full shift toward numbers in the conduct of society did not come about until the sixteenth and seventeenth centuries, reaching a real peak in the eighteenth, in the decades just before the American and French revolutions.

This transition cannot be attributed exclusively to purely scientific motives. The mathematical science of probability, for example, came into being because of the desire to know the odds in various games of chance. One of the early landmarks in the development of probability theory, *The Doctrine of Chances* (1718) by Abraham de Moivre (1667–1754), begins by apologizing for the ignoble origins of this branch of mathematics. Many readers will have encountered numbers in reckoning the odds in various games of chance. Gamblers know, for example, that a throw of the dice will yield a 7 more frequently than a 12. The reason is that there are six ways to get a seven (6-1, 5-2, 3-4/4-3, 2-5, 1-6) but only one way (6-6) to get a 12.

Those who bet on horses or dogs must be aware of the odds. Inevitably, the improbable will sometimes occur, as when the 2002 Belmont Stakes was won by a horse with odds of 70 to 1. Poker players know the odds of drawing a card to fill a straight or a flush. State lotteries attract vast numbers of bettors who have only a minuscule probability of winning. But even those who do not patronize the lotteries are made aware of spectacular wins which merit display on the front page of the daily newspapers. After all, who would not be thrilled to learn how an ordinary person became a multi-millionaire overnight?

Lotteries have been a source of income to states and institutions for hundreds of years. In the eighteenth century, Harvard College financed the building of a new dormitory by a lottery, and also used the proceeds of a lottery to raise funds for the purchase of a large orrery.

During the centuries since lotteries were first introduced, many men and women have dreamed of ways to "beat" the lottery. These schemes have generally proved to be chimerical. But several hundred years ago, one man actually did beat the system.

The architect of this triumph was Voltaire (1694–1778). Voltaire is famous for his satirical wit, his poems and plays, his historical works, and his espousal of good causes. He also wrote an introduction to Newtonian science for nonspecialists, one which remains to this day the best such work ever written. It is not generally known, however, that Voltaire was skilled in numbers and was successful at "beating the lottery." In 1728, the city of Paris defaulted on a large number of municipal bonds. The embarrassed French government set out to make at least partial restitution to some of the bondholders by subsidizing a series of monthly lotteries. Each month, holders of the defaulted bonds—one of whom was Voltaire—could buy lottery tickets at a cost of one franc per 1,000 francs of bond face value. The resulting lottery would give a few lucky winners partial restitution of their investment. Voltaire promptly saw that because of the subsidy, the payout on these lotteries would be far greater than the collective cost of buying all the tickets. Working with the French mathematician La Condamine, (1701–1774), he created a syndicate of former bondholders capable of buying all the available tickets, thereby assuring a win every month. The syndicate then distributed the winnings among the participants far more advantageously than the government would have. The eligible bondholders were delighted to exchange certainty for risk. The syndicate won the lottery every month without fail[11] Scholars differ on exactly how much money Voltaire made and whether he actually became a millionaire, but they agree that he did gain a measure of financial independence from this exercise. And he had enough good sense never to gamble on the lottery afterward.

Even in the sciences, the use of numbers as a test of the validity of a theory came into being only in the seventeenth century. Of course, ever since antiquity, astronomers and astrologers had used numbers to specify positions of celestial bodies and numerical coordinates determined locations on maps. The test of the validity of theories, however, did not come from numerical tests of prediction and observation. Of greater importance than numerical predictions and observations was the degree of harmony with the principles of Aristotle. Even such a founding work as Galileo's *Two New Sciences* (1638), the foundation of our modern system of analytical mechanics, did not end up by producing numerical examples to test the validity of the system of motion there presented. Of course, Galileo's goal was to establish laws of motion that were in harmony with nature and he took delight in showing that his rules for a naturally accelerated motion yielded numbers that conformed with the results of simple experiments. However, he did not make use of extensive numerical data from experiment and observation in the sense that Newton did some decades later in his *Principia*.

It was only with a later generation of scientists such as Christiaan Huygens (1629–1695) and Isaac Newton (1642–1727) that theory and data of observation became so closely entwined that the validity of the system lay in its success in numerical prediction.

Our mission from here on is to explicate some of the important and interesting steps whereby the analysis of society, the conduct of government, the regulation of daily life, and the understanding of nature came to be.

2 NEW WORLDS BASED ON NUMBERS

The seventeenth century witnessed the birth of modern science as we know it today. This science was something new, based on a direct confrontation of nature by experiment and observation. But there was another feature of the new science—a dependence on numbers, on real numbers of actual experience.

The pioneering practitioners of the new science knew that they were producing a new kind of knowledge and so they declared this newness in the titles of their books and articles. Thus we have Galileo's *Two New Sciences*, Boyle's *New Experiments*, Kepler's *New Astronomy*, and Tartaglia's *New Science*. When, in 1620, Ben Jonson presented a masque entitled "News from the New World," his new world was not the newly found continent of North America, but the new world of science, the world revealed by the telescope of Galileo.

A numerical law is an exact statement that leads to prediction and test. For example, one of the laws discovered in the seventeenth century, in the age of the Scientific Revolution, was Hooke's law, "Ut tensio, sic vis," a statement concerning springs. It tells us that the tension in a spring is proportional to the stretching force. Let us suppose that we hang a spring from a hook and that the spring has a pan on which we can place various weights. Then Hooke's law tells us that if the spring is stretched out by two inches by a weight of three pounds, then a weight of six pounds will stretch out the spring by four inches. At once the law can be tested. Such a test, implying the possibility of falsifying a theory, is a sign of modern science. According to Karl Popper, perhaps the foremost philosopher of science of the twentieth century, the property that a theory can be falsified by experiment is the very hallmark of science.

The ancients knew a few such numerical laws: the law of reflection of light, the law of the lever, and the law of buoyancy. But prior to the Scientific Revolution, the goal of science (or the study of nature) was not to seek laws of nature expressed in terms of numbers or number-relations. Those who created the new science of the seventeenth century not only found laws based on numbers but they were also willing to express these laws in terms of higher powers of numbers—squares and cubes.

KEPLER'S HARMONIC LAW

Typical of the new science is the discovery by the astronomer Johannes Kepler (1571–1630) of what we know today as Kepler's harmonic law. Kepler had been searching for a law that would explain why God, in creating the world according to the Copernican system, placed the planets where they are and caused them to move with the speeds that we can observe. Kepler sought in vain for many years to find some numerical law that would express the relationship between the celestial dimensions or distances and the speeds with which the planets move. Finally he found the answer. In his 1619 book *The Harmony of the World* he tells us that he discovered the harmonic law while delivering a lecture on astronomy to his students. Kepler found that for each planet, the cube of the average distance from the sun is proportional to the square of the period of revolution.

Kepler later found a similar law for the satellites of Jupiter. Today we know that such a harmonic law holds for any system of bodies that circulates around a central parent body. There are many applications of Kepler's law; for instance, half a century later it gave Isaac Newton the clue to his discovery of the law of universal gravity.

GALILEO AND THE LAWS OF MOTION

According to Aristotle (384–322 B.C.E.), "To be ignorant of motion is to be ignorant of nature." And indeed, studies of motion have been a fundamental part of thought throughout antiquity and during the Middle Ages. This motion, however, was not the kind of motion with which Galileo (1564–1642) was concerned and which we think of whenever we encounter the term "motion" today. The pre-Galilean thinkers were rather concerned with motion in the sense used by Aristotle. For them "motion" was any process in which there was a transition from any state or condition to another state. Thus the process of aging, the change in a person's degree of wisdom, or the growth in weight of a boy could all be considered examples of motion. By contrast Galileo was concerned with *physical* motion, motion involving a change in place—the set of phenomena which we today commonly associate with the term. One of the major kinds of motion that Galileo studied was the motion of free fall, the falling motion of bodies that we can observe around us, right here on Earth.

In his founding treatise, the *Dialogues Concerning Two New Sciences* (1914; originally published 1638), Galileo boasted that he was setting forth "a very new science dealing with a very ancient subject." He was aware that some superficial observations had been made, such as that a body in free fall is continuously accelerated, but nobody before him had discovered the laws of such acceleration, as he had done. No one before him, he declared, had discovered that "the distances traversed, during [successive] equal intervals of time, by a body falling from rest, stand to one another in the same ratio as the odd numbers beginning with unity [one]."[1]

A numerical example will help the reader to understand Galileo's statement. Suppose that an object is let fall and that during the first second of time, it will fall through 16 feet. Then, according to Galileo, in the next second of time, it will fall through 3 × 16 feet; during the third second it will fall through 5 × 16 feet, and so on.

Galileo's rule can be expressed differently, that the total distance fallen is proportional to the square of the total elapsed time. Thus, if a body falls freely through 16 feet in the first second, it will fall through 2^2 × 16 feet or 4 × 16 feet during the first two seconds, through 9 × 16 feet during the first three seconds, and so on.

The question then arises whether Galileo's laws hold in nature or are merely the result of an abstract mathematical exercise. Galileo could not make an experimental test with freely falling bodies, in part because they move too swiftly. So he devised an experiment in which he "diluted" gravity, slowing down the motion of falling. For this purpose he used an inclined plane, a slanting flat board with a grooved track running down its length. He allowed a small metal ball to roll down the board at different inclinations and recorded the distance and times. In every case he found that the distances, starting from rest, were proportional to the square of the time.

In his own words:

We repeated this experiment more than once in order to measure the time with an accuracy such that the deviation between two observations never exceeded one-tenth of a pulse-beat. Having performed this operation and having assured ourselves of its reliability, we now rolled the ball only one-quarter the length of the channel; and having measured the time of its descent, we found it precisely one-half of the former.[2]

Galileo presented the numerical values that he found in his experiments as proof that the laws of motion he had found were not abstract laws but actually occur in the physical world of nature. Thus he could proudly boast of an agreement to within "one-tenth of a pulse-beat."

In the next generation, that of Newton's *Principia* (1687), numerical data from experiment and observations were used in another way, to set requirements that revised or extended fundamental theory. An example is Newton's development of perturbation theory in his attempts to deal with new and more precise observations. In the *Principia*, Newton showed how fundamental theory must be modified so as to account for the numbers given by experiment and observation. Two notable examples are the numbers representing resistance to motion and the motion of the Moon.

NUMBERS IN THE LIFE SCIENCES: DOES THE BLOOD CIRCULATE?

One of the tremendous advances in science made during the Scientific Revolution of the seventeenth century was William Harvey's discovery of the circulation of the blood. Harvey (1578–1657) was a British physician who had been trained in Italy. In fact he was studying medicine at the University of Padua in the years when Galileo was using the newly invented telescope to revolutionize the science of astronomy.

In the early part of the seventeenth century, the age of Galileo and Kepler, the reigning concepts of human and animal physiology were still the ones set forth by Galen (c. 130–c. 200) in the second century. Galen believed that blood was manufactured continuously by the liver and then spread through the whole body with an ebb and flow motion like that of the tides in the ocean.

Harvey's radical concept was that the heart acts like a pump. He showed by careful anatomical and physiological "exercises," and by comparative data from various kinds of animals, that the heart pumps blood out into the aorta or primary artery; with each successive heartbeat the blood is pushed further and further out through the system of arteries; eventually this blood returns to the heart through the veins. In making this analysis, Harvey used the discovery of valves in the veins. These valves, as Harvey showed, permit the blood in the veins to move toward the heart but not away from the heart. In other words, the blood could not possibly ebb and flow in the system of veins.

In Harvey's concept, basically the one in which we still believe today, the blood that returns to the heart through veins is not then directly pumped into the main system of arteries. Instead, this returning blood is pumped into the lungs, where it is aerated and then returns to the heart, where it is pumped once again into the aorta and the main system of arteries. One of Harvey's telling refutations of the Galenic system, and the reason why Harvey appears in a book about numbers, is that he used an argument based on numerical evidence. Thus Harvey's discovery shows us that the life sciences as well as the physical sciences made some use of numerical arguments.

Harvey's quantitative argument against the Galenic system was based on two numerical quantities. The first was the capacity of the heart in human beings, in dogs, and in sheep. Then, multiplying this figure by the pulse rate, he computed how much blood is transferred from the heart—approximately 80 pounds of blood in each half hour for an average man. From these measures, Harvey wrote, it is manifest "that the beating of the heart is continuously driving through that organ more blood than the ingested food can supply, or than all the veins together at any given time can contain."[3]

The critical reader will find two flaws in Harvey's argument. The first is that he supposed the average human pulse rate to be 33 whereas the average or normal pulse is closer to 70 beats per minute. As a consequence, his calculations were off by a factor of 2. The second is that it is very difficult to measure the capacity of the heart. Harvey obtained a human heart from an anatomized cadaver and simply poured in water, measuring the amount of water that the heart could contain. But under these conditions the walls of the heart are much more flaccid than in the sturdy conditions of life. In fact, scientists did not make accurate measures of the heart's capacity until well into the nineteenth century. Nonetheless, Harvey's numerical argument proved the falsity of Galen's basic postulates: There is simply no way in which the liver can continuously manufacture the quantity of blood being pumped out by the heart.

At more or less the same time that Harvey was studying the circulation, other scientists were applying numerical considerations to the life sciences. For example, Johannes van Helmont (1579–1644) of Belgium made quantitative studies of the nourishment of growing plants. In addition, Italian scientist Santorio Santorio, also known as Sanctorius (1561–1636), performed studies to determine the correlation of food intake and body weight. Sanctorius had a platform constructed which was supported by slings in such a way that it could be weighed while he sat on it (see figure 2.1). He then made careful measurements of his weight at regular intervals during a whole day, and thus was able to correlate the changes in his weight with the weight of solid food and liquids he ingested. He also weighed his solid and liquid excreta, and by numerical methods established the weight loss due to perspiration.

In assessing the novelty of a science based on numerical laws

FIGURE 2.1 Sanctorius in his weighing chair, from his 1614 *Medicina statica. Courtesy of the National Library of Medicine*

such as those of Galileo, Kepler, Harvey, and Sanctorius, we must keep in mind that although it was new to express laws of nature in terms of numbers, by the time of the Scientific Revolution, numbers had been used for centuries in various aspects of life. Surveyors and tax collectors had obviously been concerned with numbers. Jewelers and dealers in precious stones and metals had used numbers and measured their wares by use of a balance. Apothecaries had also used numerical weights in measuring medicines. Astronomers had measured stellar and planetary positions (altitude and azimuth or Right Ascension and declination) since at least the second century. Mapmakers had for centuries been locating positions of cities and natural features of the landscape in terms of two numerical quantities: longitude and latitude. And of course, astrologers had been making numerical calculations as part of their casting of horoscopes. But before the age of Kepler, Galileo, and Harvey, numbers were not used to express general laws of nature or to provide testable questions to test a scientific theory. This feature of the use of numbers in science set the new science of the Scientific Revolution apart from the traditional study of nature; in fact it defines the newness of the new science.

The numerical character of the new science also appears in the invention of new instruments for measuring physical quantities. Among these are the pendulum clock for measuring time and the barometer for measuring air pressure. One of the great innovations of the seventeenth century was the invention of the telescope. Historians have explored how this new instrument changed our ideas about the Moon and the planets, and revealed the existence of new stars never before known. Less attention has been paid to the significant new level of precision attained when telescopes became equipped with micrometers. The telescope showed what planets are like and revealed details such as the mountains on the Moon, but it was only with the introduction of the micrometer that measurements could be made. This transformation of the telescope shifted the focus of astronomy from qualitative description to quantitative measurement.

A FIRST EXERCISE IN DEMOGRAPHY: HOW MANY PEOPLE CAN THE EARTH SUPPORT?

Many readers will be acquainted with the name of Antoni Van Leeuwenhoek (1632–1723). Leeuwenhoek is generally considered to be the father of microscopy because his pioneering discoveries established the microscope as a primary tool of biological science. He was by profession a draper who spent his life in the town of Delft in Holland. He was not educated in science, but he was a skilled artisan and instrument maker who constructed his own microscopes. (He left no records of his mode of producing the lenses that permitted high magnification.) Leeuwenhoek had an uncanny faculty of knowing what to look at, that is, what subjects would yield important information concerning different aspects of the life process. He also knew where to send his reports on his new discoveries—to the Royal Society in London, at that time the foremost scientific society in the world. They published his reports and honored him for his research by electing him a fellow of the Royal Society.

Leeuwenhoek was not particularly interested in demography, in problems of the world population. He came to this area because he was searching for a means of expressing in numbers the size of the objects he had been studying with his microscope. Leeuwenhoek saw that there are a number of ways of indicating quantity, as we may see by looking at cookbooks. A recipe might call for a "pinch" of salt, a tablespoon of this, or a teaspoon of that. Or, the recipe may call for half a pound of butter. Another way of indicating size would be by volume or area. Thus we might indicate the size of a Ping-Pong ball by the number of them that could be placed in a gallon pail.

Leeuwenhoek was concerned with the problem of size in relation to one of his most important discoveries: the spermatozoa. He wanted to give his readers some idea of the minuteness of these "animalcula." On 25 April 1679, he wrote to the Royal Society concerning his discovery. His communication announced that the number of "little animals in the milt of a cod" (150 billion) was far greater than the total number of people that the earth could support.[4]

How did Leeuwenhoek compute the number of people the earth could support? In order to achieve this result, he did not resort to geometry, the mathematics of the university. Instead, he used arithmetic, a subject which he knew as a businessman, the arithmetic of shopkeepers.

Leeuwenhoek's determination of the maximum number of people on the earth began with an estimate of the extent of the earth's surface. He came up with 9,276,218 square miles. At that time the Dutch mile was reckoned at one-fifteenth of a degree or about 7.4 kilometers. He assumed that vast oceans occupied two-thirds of the total surface of the earth. He also estimated that two-thirds of the dry land was inhabited. Thus the inhabited part of the earth occupied 2,061,382 square miles. If we know how many people a square mile can support, then it is a simple job of multiplication to compute the maximum number of people that can live together on the earth.

To find out how many people could live in a square mile, Leeuwenhoek turned to the part of Holland he knew best. This was a region comprising North- and South-Holland and part of Brabant. This region is roughly in the shape of a rectangle whose area is 154 square miles. Holland did not have a national census until well into the next century, so Leeuwenhoek made use of the fact that a head tax or "capitation" had been levied in 1622. From this he could say that this area contained a population of about a million persons.

Leeuwenhoek was now in a position to determine the size of

the maximum population of the earth. He said, let us "assume that the inhabited part of the earth is as densely populated as Holland." He was of course aware that "it cannot well be so inhabited." Since the inhabited part of the earth is "13,385 times larger than Holland," the result is a maximum of "13,385,000,000 human beings on the earth." In the present context, the way in which Leeuwenhoek arrived at his result is not of much interest. What is significant is that Leeuwenhoek's desire to express in numbers the minuscule size of the animalcules he had discovered led him to produce the first recorded exercise in the numerical science of demography.

THE NEED FOR LIFE TABLES

In the seventeenth century there arose a new and important industry using numbers—life insurance. A person would purchase an annuity by making payments to an insurer, and then after a certain date, the insured person would receive regular payments from the insurer for the rest of his life. The specific terms of purchase were agreed upon for each annuity. Both the insured and the insurer would want to get the most advantageous terms possible, and for this purpose they needed to know the average life expectancy of individuals in the insured's age category. Of course, there was no way of predicting whether the individual insured would fit the average. Some people frowned on this kind of business because it was a form of gambling on a person's life.[5]

The problem of supplying numerical tables for use in computing annuities was taken up by Edmund Halley (1656–1742), the astronomer after whom the comet is named, a skilled mathematician, and secretary of the Royal Society.

Halley began by ascertaining the number of burials in Paris and London in 1680: 24,441 in Paris and fewer than 20,000 in London. He found that the numbers of births and marriages were in the same proportion in each city.[6] Such data, however, were subject to fluctuations, because of people moving between the country and the city. To supply reliable tables, Halley needed numbers for a stable population.

And so we can imagine the joy with which Halley would have received tables of the births, marriages, and deaths recorded in the town of Wroclaw (Breslau) in Silesia, now in Poland. These data had been compiled by Caspar Neumann (1648–1715), a pastor and an ecclesiastical judge in the reformed church. Believing that his data were important, Neumann sent them to Leibniz, the outstanding intellectual of the German-speaking world at the time. Leibniz, seeing no special value to these data, sent the figures on to Paris, where again their value was not recognized. A French correspondent of the Royal Society sent the compilation on to London, where it came into the hands of the secretary, Halley. Halley at once recognized that here were just the numbers he needed, based on careful records kept for a stable population. Few people moved to Wroclaw and few people moved away.

Halley used Neumann's numbers as a basis for a memoir which is celebrated as having laid the foundations of a correct theory of the value of life annuities. The memoir was published in the *Philosophical Transactions of the Royal Society* for 1693. The memoir bore the title "An Estimate of the Degrees of the Mortality of Mankind, drawn from curious Tables of the Births and Funerals at the City of Breslaw; with an Attempt to ascertain the Price of Annuities upon Lives." Table 2.1 is a partial summary of the data.

A NEW WORLD OF NUMBERS

In recent decades, a group of scholars has been exploring how numerical considerations brought into being a science of statecraft and a mode of social analysis worthy of the name *social science*. Their endeavors have shown how quantitative considerations have

TABLE 2.1 SUMMARY OF DEMOGRAPHIC DATA FROM BRESLAU, 1687–1693, WITH ADDITIONAL DATA FROM NEUMANN FOR 1694

Year of Birth	Number Born	Number Surviving in the Year						
		1688	1689	1690	1691	1692	1693	1694
1687	1,186	940	884	792	738	708	680	662
1688	1,214		956	880	761	714	685	663
1689	1,191			954	849	743	690	655
1690	1,312				991	903	800	723
1691	1,292					988	890	759
1692	1,151						857	756
1693								912

From E. Halley, "An Estimate of the Degrees of the Mortality of Mankind," Philosophical Transactions of the Royal Society of London 17 (1693), pp. 596–610.

entered the conduct of government and have given us a new level of understanding of social systems. Thanks to the research of these scholars, we are now aware of the stages in the creation of a new science—statistics—that has affected almost every aspect of our lives.[7] For most readers, the word "statistics" will suggest a table of numbers, such as the data relating to the performance of members of a baseball team. This concept of statistics will seem quite proper since the singular "statistic" means a single numerical datum. The singular form, "statistic," seems to have come into general usage only in fairly recent times, whereas the more general word "statistics" came into being in the seventeenth and eighteenth centuries. Furthermore, the noun "statistics" is singular rather than plural.

A British historian, Keith Thomas, and his American counterpart, John Brewer, have carefully traced the increasing dependence of government on numbers for sixteenth- and seventeenth-century England. They find that during the seventeenth century the expanding economy of England and the problems of military statecraft created a pressing need for numerical information in different departments of government. Ministers of the Crown needed quantitative information on "all of the various resources of the different departments [of government] in order to exercise firm control over government policy."[8] The Parliament, "both as a policy maker and as the body dedicated to securing a responsible executive," needed government statistics. Such statistics were also needed by various "occupational groups and special interests directly affected by state policies." They were "eager to learn the grounds" on which decisions were being made. Additionally, according to Brewer, the general public had developed "a substantial appetite for the sorts of [numerical] information that only the very considerable resources of the state could provide."

A striking example of the use of numbers in relation to the conduct of life is the regular tabulation and publication of Bills of Mortality. Such bills seem to have originated in outbreaks of the plague, when there was a great concern to know whether conditions were getting better or worse. The King, members of the Royal Court, and ordinary citizens wanted to learn whether it was safe to remain in town or whether to seek refuge in the country. Toward this end regular statements were published concerning the number of deaths. These were called Bills of Mortality. In London, following the plague epidemic of 1603, these Bills of Mortality came to be produced and published regularly, even during times of no epidemic. These broadsides listed the number of deaths in the week just past. According to John Graunt, who in the seventeenth century became the first person to analyze these Bills of Mortality, special attention was paid to "how the *Burials* increased, or decreased," and "among the *Casualties*, what had happened rare, and extraordinary in the week current." These bills were especially important in "the *Plague-time*," as indicators of "how the *Sickness* increased, or decreased, so that the *Rich* might judge of the necessity of their removall" from London and so that "*Trades-men* might conjecture what doings they were like to have in their respective dealings."[9] These numerical tables were of special significance because they provided the database for what appears to have been the first truly statistical analysis in recorded history.

Historians generally agree that the first example of a statistical analysis, as we understand this term today, came in Graunt's 90-page book entitled *Natural and Political Observations Mentioned in a Following Index, and Made Upon the Bills of Mortality* (London, 1662). Keep in mind that Graunt was not a university professor, nor even a college-educated amateur scientist, but rather a merchant engaged in a small business. He is said to have been a draper (a seller of cloth) or a haberdasher.

We learn about Graunt's life through accounts of him by a contemporary, John Aubrey (1626–1697), who collected information concerning the lives and achievements of famous men he had known or heard about. From Aubrey we learn that as a young man Graunt customarily arose very early so as to spend several hours working up a good reading knowledge of Latin and French, but Aubrey does not tell us why Graunt wanted to learn these languages.

Graunt evidently became an important member of the London business community, active in civic affairs. He became an officer in an organization known as the "Trayned Band," and Aubrey refers to him as "Captaine John Graunt" (afterward, major). His intellectual distinction was recognized in 1663 when he was elected a fellow of the Royal Society. According to Aubrey, Graunt was apparently a person of some influence; he obtained for his friend William Petty the post of professor of music in Gresham College and also a lucrative appointment to be "one of the Surveyors of Ireland."[10]

Graunt's business was wiped out by the Great Fire of London of 1666. Perhaps as a result of despair, Graunt then converted to Roman Catholicism. We may agree that at a time when Graunt "most needed the earthly help of those able to provide it, his spiritual beliefs gave them an excuse not to do so."[11]

The path-breaking significance of Graunt's contribution to knowledge was at once recognized by his contemporaries, and it was on the basis of his little book that he was elected a fellow of the Royal Society. Yet Graunt did not become a regular attendee at the weekly meetings of the society and he made no further contributions to knowledge; he even let his membership in the society lapse. He did, however, produce several expanded and revised editions of his book.

Graunt's analyses were based, as the title of his book declares, on the London Bills of Mortality. He begins with a brief history of these bills. Until 1629, only summaries or totals were provided. Then the totals of christenings and burials were divided according to sex and the causes of death were given. In the bills for 1632, deaths were classified according to 63 different causes. For example, the 9,535 deaths listed for 1632 begin with 415 "Abortive and Stillborn" cases, one case of "Affrighted" death, and 628 "Aged." There were 2,268 deaths of "Chrisomes" (newly baptized infants) and other infants, plus 267 deaths from "Dropsie and Swelling," 1,108 from fever, and nine from "Scurvy, and Itch."[12] Graunt also makes the point that he was the first person to analyze these data and that the diagnosis of the cause of death was made by ignorant enumerators. As Graunt studied this collection of numerical data he was struck by the statistical regularities that appeared. Graunt found that "among the several *Casualties* some bear a constant proportion unto the whole number of *Burials*." These included chronic diseases, accidents, and suicides. But he found that "*Epidemical* and *Malignant* Diseases," such as the Plague, "do not keep that equality, so as in some Years, or Moneths [*sic*], there died ten times as many as in others."[13] One of his most interesting findings is that slightly more males than females are born each year.

One of the celebrated passages in Graunt's book is his account of the way in which he used the numerical data in the Bills of Mortality to arrive at an estimate of the size of the population of London. In this exercise, he showed how to interpret these numbers, how to draw inferences from them. Graunt's estimation of the size of the population of London is an outstanding example of statistical analysis. It shows the way in which tables of numbers can be made to yield statistical information. In other words, this is a pioneering attempt to interpret numerical data and cause them to yield a reasonable estimate of a significant social number.

Graunt's computation of the London population draws the admiration of everyone who studies the history of statistics. Others before him had made guesses concerning the magnitude of such populations, but Graunt was the first person to base his figure on reasonable inferences from actual numerical data.

As his base figure, Graunt used the annual number of christenings as listed in the Bills of Mortality. This number, he pointed out, was subject to some error. It was arrived at from the entries in parish registers and so did not include the small number of Catholics, freethinkers, or other religious dissenters. Furthermore, there was no way of knowing the number of parents who, though nominally members of the established Church, had not had their children baptized.

Graunt's first stage of computation was based on the conclusion that, on average, women of childbearing age would give birth to a child every other year. Or, as he put it, such "women, one with another, have scarce more than one Childe in two years."[14] Since there were on average some 12,000 children born in London each year, Graunt concluded that the total number of "Teeming women" (or women producing children) in London would be 2 × 12,000, or 24,000.

Next, he ascertained that the time span during which women gave birth extended over 24 years, from age 16 to age 40. He then took note that the number of women who were married ranged in ages from 16 to 76 years. Thus, the total number of women living in London would be about twice the number of women of childbearing age, or 2 × 24,000 = 48,000. Of course, Graunt's 16 to 76 is a span of 60 years, which is somewhat more than 2 × 24. But Graunt was not attempting to produce an exact number so much as an order of magnitude.

The final number was based on another finding of Graunt's. He found that the average family unit in London tended to consist of eight individuals: a mother and father, three children, and three servants or lodgers. Thus the total population of London would come out to be 8 × 48,000, or 384,000.

Is this a credible number? It's a big number, but is it big enough? Graunt found two ways of checking the validity of his procedure. The first was to consider the actual number of inhabitants in certain parishes. In these, he found, there were three deaths annually per 11 families. He also knew that the total number of deaths in London was some 13,000 per annum. Hence, he concluded, the number of families in London must be $(11/3) \times 13,000 = 47,700$, that is, about 48,000 families. This is the same number of families that he had arrived at by his earlier calculation and so it gave him confidence in his method and his concluding number.

Graunt's second check on his procedure was based on the map of London. He assumed that a reasonable estimate of the total number of families living within the town walls was 12,000. He also found from the Bills of Mortality that the number of deaths outside the walls was three times the number of deaths within the walls. Hence, once again, he came out with 48,000 family units for London as a whole, a quarter of which lived within the town walls and three-quarters of which lived in the surrounding area.

A modern appraisal of Graunt's results is that "they were perhaps not very far from the truth." Graunt's mortality rate of 3 out of 88 (corresponding to his eight members of the 11 families) would be expressed in our times as a rate of 34 per 1,000, which "is not improbable."[15] Of course, Graunt made some bold assumptions with which we can easily quarrel. But what is most important is not whether his work would pass muster today. Rather, what is historically significant is that he showed by example how numerical data could provide a basis for interpretation; he demonstrated how a statistical point of view could elicit conclusions of general interest from tables of specific numerical data.[16]

SIR WILLIAM PETTY AND POLITICAL ARITHMETIC

It would be difficult to find a pair as different in lifestyle, career, and influence as Graunt and Petty. Whereas Graunt was shy and retiring, dying in poverty and bankruptcy, Petty was a swashbuckling character, an adventurer, a confidant of the ruling monarch Charles II. He was a bold investor who amassed a fortune and died a wealthy man.

Sir William Petty (1623–1687) was in a real sense Graunt's intellectual heir and successor. Whereas Graunt wrote a single work, Petty authored enough books to justify one written about his results by Sir Geoffrey Keynes, the brother of the economist.[17] Petty's multifaceted career has been summed up as follows. He appears, Keynes wrote, "as cabin boy on an English merchant ship, as peddler of sham jewelry in France, as naval cadet, as medical student in the Netherlands and in Paris, as reader in music in Gresham College, as Professor of anatomy at Oxford, as Fellow of the Royal Society, as educationist, inventor, Latin versifier, and ship builder, as physician to Cromwell's army in Ireland, as surveyor and geographer of Ireland, and in the midst of all that as a man becoming ever more involved in administration, finance, and politics."[18]

Petty was born in 1623 in Romsey, a rural community, where his father was a clothier who dyed his own cloth. Young Petty was fascinated by the work of artisans—smiths, watchmakers, carpenters, and so on. After an apparently sound schooling, Petty—at the age of 14—went to sea as a cabin boy. While cruising in the Mediterranean, Petty broke his leg and was put ashore at Caen, unable to get around and ignorant of the French language. How he managed it, we do not know for sure, but Petty healed and became fluent in French. He also improved his Latin. He evidently studied mathematics and navigation and when at the age of 20 he returned to England, he joined the Royal Navy.

Petty's career in the Navy was short-lived, and in 1643 he returned to the Continent where he studied medicine and anatomy in the Netherlands and in Paris. He was so poor during these student years that, as he told John Aubrey, he had to live for a whole week on three pennies' worth of walnuts. Back in England, Petty eventually (in 1649) became a Doctor of Medicine.

Somehow or other, Petty made the acquaintance of Graunt, who recommended him for a post in music at Gresham College, London. Later Petty became a professor of anatomy at Oxford and eventually was elected Master of Brasenose College in Oxford. Petty was interested in the new science coming into being during the Scientific Revolution and became one of the founders of the Royal Society of London.

At Oxford Petty became well known for his daring act of reviving a woman executed by hanging. She had been found guilty of murdering her illegitimate child. Her body was turned over to Petty to be used as a subject in the dissecting room. Petty was able to revive her, although she had been pronounced dead. Evidently she had been "inefficiently hanged." Petty reported this medical phenomenon as "History of the Magdalen" and it was published in the popular press as *News from the Dead* in 1651.[19]

In 1652 Petty accompanied Cromwell's troops into Ireland. He evidently didn't serve in the simple capacity of doctor to the army, but rather as a surveyor, busy with evaluating the Irish estates that were being taken over by the British. Petty acquired land and leases for himself and so ended up with a minor fortune. Despite his association with Cromwell and his armies, Petty became a favorite of King Charles II after the Restoration. Perhaps Charles liked Petty's wit, but it has also been suggested that he was grateful to Petty for having uncovered new sources of tax revenue to support the royal lifestyle. In any event, Charles thought well enough of Petty to grant him a knighthood in 1662.

From the point of view of the history of numbers, Petty is important for his early attempts to determine the population of England and of Ireland. Recognizing the significance of Graunt's work, Petty advocated that "policy"—that is, the conduct of government—should be based on numerical data. Petty developed a theory of social and economic understanding and planning that he named "political arithmetic." In fact, Petty wrote and published five works with the words "political arithmetic" in the title. In one of the most important of these, simply called *Political Arithmetick* (1690), he described this new form of statecraft as follows:

> The Method I take to do this, is not yet very usual; for instead of using only comparative and superlative Words, and intellectual Arguments, I have taken the course (as a Specimen of the Political Arithmetick I have long aimed at) to express my self in Terms of *Number*, *Weight*, or *Measure*; to use only Arguments of Sense, and to consider only such Causes, as have visible Foundations in nature; leaving those that depend upon the mutable Minds, Opinions, Appetites, and Passions of particular Men, to the Consideration of others.[20]

It should be noted that Petty not only had a vision of a new statecraft based on numbers but also declared that the mathematics of numbers, that is, algebra, was the tool for making the analyses on which the new statecraft would depend.

"Political arithmetic," as we shall see, became widely used during the eighteenth century to denote a statistically based statecraft—one based on numerical information concerning such entities as population or demography, natural resources, manufactures, exports and imports, and agriculture. The term fell into disuse in the nineteenth century. We may note that this subject is only imperfectly described as an arithmetic. The science of statistics, it has been observed, requires the mathematics of probability; arithmetic and even algebra do not suffice.[21]

Yet, whatever disciplines it encompassed, Sir William Petty's "political arithmetic" was moving down a rational path that would not be followed by everyone.

3 NUMEROLOGY AND MYSTIC PHILOSOPHY: SCIENTISTS AT PLAY WITH NUMBERS

We think highly of the seventeenth-century pioneers who introduced numbers into the study of nature and the analysis of society. However, the seventeenth century was also a time of flourishing interest in traditional numerology and in number-based mystical philosophy. Many people believed that numbers provided a key to human fate—some numbers are good, others bad. For example, 7 was (and by many, still is) considered a favorable number, whereas 13 tends to be associated with bad luck. The particular combination of weekday and day of the month in Friday the Thirteenth was then (as it is now) often considered an unfavorable day, a time of danger. Various explanations are given for the belief that Friday is unlucky: that Jesus was crucified on a Friday, that Adam and Eve ate of the forbidden fruit on a Friday, or that the Great Flood began on a Friday.

For the Japanese, 4 is an unlucky number, because in the Japanese language the word for "4" sounds like the word for "to die." As a result, in Japan no items such as teacups or fruit are sold in sets of four, but rather in sets of five.

Among various religions, 33 has long been a special number. For Christians it is the number of years that Christ lived on Earth. For Jews familiar with their Old Testament, 33 is the number of years of David's reign. Muslims know 33 because prayer beads commonly contain three groups of 11 beads separated by a "witness." These reverential associations sometimes intersect with more rational considerations.

Isaac Newton, whose *Principia* is generally considered to be the high point of the Scientific Revolution, was fascinated by numbers. He devoted much time and energy to computing the magnitude of the "sacred Cubit of the Jews," and he also wrote three works on his computations of the size and structure of the Temple of Solomon.[1] He left no clue as to what these numbers represented, but most likely these were associated with his interest in prophecy. Newton's biographer, R. S. Westfall, has suggested that it was in connection with these studies of Solomon's temple and the numbers relating to the temple's size and structure that Newton undertook the study of Hebrew.[2]

WHAT'S IN A NAME?
CONVERTING NAMES INTO NUMBERS

The Book of Revelation according to Saint John, along with the Book of Daniel, abounds in references to numbers. There are "seven angels" that held "seven trumpets." A "third of the sea became blood." There are "twenty-four thrones, and seated on the thrones were twenty-four elders." In the Book of Revelation numbers are especially associated with the "Beast" that blasphemes God; the Beast is said to have "ten horns and seven heads."

In Rev. 13:18 we learn that numbers represent a man's name and that the numerical value of the letters in the name of the Beast is six hundred sixty-six. Annemarie Schimmel, a scholar who has devoted a lifetime to the study of numbers and their uses, observes that the number 666 "has nourished the imagination of generations of Christians and is still much discussed today."[3]

TABLE 3.1 A NUMEROLOGICAL SCHEME BY PETRUS BUNGUS

A	1	K	10	T	100
B	2	L	20	U or V	200
C	3	M	30	X	300
D	4	N	40	Y	400
E	5	O	50	Z	500
F	6	P	60		
G	7	Q	70		
H	8	R	80		
I or J	9	S	90		

From P. Bungus, Numerorum mysteria

Two numerologists in particular have exercised their fantasies in converting names into numbers. Petrus Bungus (b.?–1601) authored a popular numerological treatise *The Mystery of Numbers*, first published in 1585. This book aroused sufficient attention to warrant new editions in 1591 and 1618.[4]

In Bungus's system, numbers were assigned to letters according to the scheme shown in table 3.1. Note that I and J were the same letter, as were U and V; there was no W.

As a good Catholic, Bungus wanted to show that if the letters in Martin Luther's name were converted into numbers, they would add up to 666, the number of the Beast. The letters in MARTIN add up as follows:

M	30
A	1
R	80
T	100
I	9
N	40

The total is 260. LUTHER is then treated the same way.

L	20
U	200
T	100
H	8
E	5
R	80

LUTHER adds up to 413. The sum of the two (260 + 413) is 673, seven greater than the goal of 666.

Like all numerologists, Bungus fiddled with numbers so as to get the desired result. His method was to Latinize Luther's surname (Luther) but not his given name (Martin). The result was:

M	A	R	T	I	N	L	U	T	E	R	A
30	1	80	100	9	40	20	200	100	5	80	1

which sums to the desired 666.

Michael Stifel or Stifelius (1487–1567) offered another numeric identification for "the Beast." Stifelius's work is particularly significant because it served as a primer to Benjamin Franklin when he was constructing "magic squares." In 1554 Stifelius published a new edition of a book on algebra, which is one of the earliest works in which "+" and "−" are used.[5] In this edition Stifelius inserted his own calculations of the number of the Beast. His target was Pope Leo X and he set out to convert that name into numbers, using the Latin form, LEO DECIMUS.

Treating the pope's name as Bungus had done with Martin Luther's, Stifelius found that the sum of the numbers was only 416, short of 666 by 250. So, in the manner of numerologists, he began fiddling. Turning away from Bungus's system, he pulled from Leo's name only the L, D, C, I, M, and V (since U = V)— the letters that are also Roman numerals. Then he took away the M on the ground that it stood for "mysterium" or mystery and so could be discarded. Finally, he added an X to represent the "ten" in Leo the Tenth. Placing the remaining Roman numerals in descending order, he got DCLXVI (500 + 100 + 50 + 10 + 5 + 1), or 666. Thus by suitable fiddling, he was able to show that the name of Leo X could be made to signify the sign of the Beast.

This "evidence" against the Papacy is said to have given him the final push to conversion from Catholicism to Protestantism. When Luther accepted Stifelius into the fold, he urged him to abandon his numerology. Stifelius, however, went on to find within the writings of the prophet Daniel the date for the end of the world. He announced that this event would occur on 3 October 1533. When the day came and the world continued, only Luther's interference protected Stifelius from the ire of his parishioners.[6]

NUMEROLOGY IN SCIENCE

One of the puzzles of the Scientific Revolution is the use of numbers in ways that we would not now consider "scientific." Such quasi-numerological oddities appear in the writings of very serious scientists.

Christiaan Huygens (1629–1695) was one of the leading figures of the Scientific Revolution and the only one really comparable with Isaac Newton. One of his remarkable achievements was the discovery that the planet Saturn, in addition to being endowed with its celebrated ring, has a satellite or moon. This was perhaps the most exciting discovery about the solar system since Galileo had found four moons moving in orbit around Jupiter.

The discovery of Saturn's moon was so astonishing that Huygens wanted to delay announcing it until he was absolutely certain that he had indeed found what he thought he had; he wanted time to confirm his discovery. Yet he wanted to be sure that he would get credit for the discovery if it turned out that Saturn really has a moon.

In Huygens's day, the way out of such a dilemma was to publish an announcement of one's discovery in the form of an anagram. The disclosure by anagram was a scientific case of having your cake while eating it. If the discovery was confirmed, then the anagram would be publicly unscrambled and Huygens would be given credit for his achievement as of the date of the publishing of the anagram—not the solution of it. On the other hand, if it turned out that Huygens had been mistaken, that he had seen a star or a comet, for example, and had thought it was a satellite of Saturn, then he would never have to reveal the content of the anagram or be known to have been wrong. This method of assuring one's priority while not fully disclosing the nature of the discovery was fairly common in the seventeenth century. In this same way, Galileo had announced that Venus exhibits phases, just as our Moon does. (He was right.)

Almost 20 years after Huygens's discovery, in 1672, Jean-Dominique Cassini (1625–1712) found three more moons of Saturn. Huygens thereupon felt called upon to explain why he had ceased to look for moons after having discovered what turned out to be only one of several.

Huygens said that, as a disciple of Descartes (1596–1650), he was interested in the properties of numbers. A property of some numbers is that they are "perfect." Here the adjective "perfect" merely means that the number is the sum of its divisors (numbers that divide it "perfectly," without a remainder). For example, six is a perfect number: its divisors are 1, 2, and 3; $1 + 2 + 3 = 6$. (The next "perfect" numbers are 28, 496, and 8,128.)

Huygens explained that in the Copernican system, there are six primary planets: Mercury, Venus, Earth, Mars, Jupiter, and Saturn. With the discovery of the Saturnian satellite, there were also six secondary planets or satellites: our Moon, the four satellites of Jupiter discovered by Galileo, and the satellite of Saturn. Since there was a perfect symmetry between the primary and secondary planets, both of the number six, Huygens explained, he had not even attempted to look for further satellites.

Cassini himself engaged in number fantasies. He had discovered the new satellites of Saturn while still in Italy and when he was hoping that King Louis XIV would appoint him to the post of director of the new Paris Observatory. In publishing his account of his discovery, he saw a way of ingratiating himself with the king. With Huygens's discovery, he wrote, there were 12 circulating bodies in the solar system: six primary and six secondary planets. But now, with the discovery of two further satellites of Saturn, the heavenly system consisted of 14 circulating bodies and so declared the glory of XIV and the monarch who was the fourteenth of his name!

A CRUSADER AGAINST NUMEROLOGICAL SUPERSTITIONS

Even so, one of the activists who sought to stem the tide of numerological superstition was a Lutheran minister, Pastor Caspar Neumann (1648–1715) of the city of Breslau, who furnished the data for Halley's life tables. Neumann based his campaign on numerical records of parishioners going back over more than a century. He used this numerical information to combat some of the number superstitions in common belief at that time. For example, he showed the fallacy of the common belief that the health of people was influenced by the phases of the Moon. Another target of his campaign was the popular superstition known as "climacteric years," that the human body is subject to a seven-year cycle of illness. Every seventh year of a person's life was believed to be a critical time for health and well-being. The ages of 49 (7×7) and 63 (9×7) were held to be especially critical years, times of great danger to health.

There was also a popular belief in "climacteric weeks." These were supposedly times when infants would be especially subject to illness and death. These dangerous times were thought to be the ages of seven weeks or nine weeks, or weeks that were multiples of seven or nine, known as the "septernarii" and the "nonarii."

Neumann was aware that the best weapon in the fight against numerology would be numerical data. Indeed, Harald Westergaard (1853–1936), a leading historian of statistics, has observed that Neumann was apparently the first person who tried to combat superstitions with statistical evidence. His arguments against climacteric years were sounder than those against climacteric weeks. As Westergaard points out, the "rate of infant mortality from one week of age to another in the first year of life varies so much that it would require better methods than Neumann's to get sound results."[7]

While Neumann was right to argue that genuine statistical evidence was the most potent weapon to fight numerological superstition, he did not live long enough to see the Age of Reason elevate this insight into a fundamental principle of public policy.

4 NUMBERS IN THE AGE OF REASON

The latter part of the eighteenth century is known as the age of the American and French revolutions. This was also the time of a growing concern with numbers, especially national numbers. In Britain the century's end witnessed the completion of Sir John Sinclair's 20-volume census of Scotland, a monument of statistical information and analysis. In the United States, the Federal Constitution required that a census be taken every 10 years.[1] This was the first time in history that a nation established a regular census. And during the decade from 1789 to 1800, the French government became increasingly aware that numerical information was crucial for statecraft. To carry on the wars that engaged the new French republic, a count of men by age categories was needed to ensure a supply of military personnel. Also, the cost of warfare demanded exact and reliable numerical data on actual and prospective tax revenues to support the war effort.

Aside from these practical applications, to which we will return in our discussion of Thomas Jefferson and Benjamin Franklin, the eighteenth century witnessed important studies on plant and animal physiology and writings on human happiness. These express an ideal of quantification, even though they do not end up in tables of numerical data.

HUTCHESON'S MORAL ARITHMETIC

One important thinker of the early eighteenth century was a Scottish philosopher, Francis Hutcheson (1694–1747). Although Hutcheson's works were widely read and frequently reprinted in the early eighteenth century, modern readers are more familiar with his contemporaries of the Scottish Enlightenment, such as the philosopher David Hume and the economist Adam Smith.

Hutcheson's first book, *An Inquiry into the Original of our Ideas of Beauty and Virtue* (1725), has been described by Garry Wills as "a sturdy piece of prose, as well as a strict logical exercise—it set the standard for expository writing that David Hume and Adam Smith, Thomas Reid and Adam Ferguson would maintain."[2] On the basis of this work, Hutcheson won a professorship of moral philosophy at the University of Glasgow, where he had an illustrious and influential career.

Here we are not particularly concerned with Hutcheson's place in the history of philosophy, but rather with his attempt to analyze the moral sense by setting up algebraic expressions. For example, in his discussion "to find a *universal Canon* to compute the *Morality* of any Actions, with all their Circumstances, when we judge of the Actions done by our selves . . . [according to] *Propositions,* or *Axioms,*"[3] he proposes the following algebraic expression:

$$B = \frac{M \pm I}{A}$$

In this equation, *B* is "*Benevolence,* or *Virtue* in any *Agent,*" *M* is the agent's "*Moment* of *publick Good,*" meaning "the *Quantity* of *publick Good* produc'd by him," *A* is "his *natural Ability,*" and *I* stands for "*private Interest.*"

Hutcheson used this algebraic relationship to translate several commonsense notions about morality into mathematical language. The first is that if two people have the same natural ability to do good (A), the one who produces more public good (M) is more benevolent (B). Conversely, if two people produce the same amount of public good, the one with more ability is *less* benevolent (since it was in that person's ability to do more). The plus/minus sign in the equation allowed Hutcheson to factor in self-interest.

Suppose, for example, that an act is good for the public and also directly benefits the person performing that act. Compare this with an act that is good for the public yet directly bothersome, or even harmful, to the person involved. According to Hutcheson's formula, the first person's benevolence is less than the second person's benevolence. In the first case, the person's self-interest offsets the public good ($M - I$); in the second case, it boosts it ($M + I$). Hutcheson did not assign actual numbers to these entities, but assumed that they could be quantified or reduced to numbers.

Another significant reason to include Hutcheson in a history of numbers is that he concluded from his algebra that "in equal *Numbers,* the *Virtue* is as the *Quantity* of the Happiness, or natural Good."[4] That is, he taught that "*Virtue* is in a *compound Ratio* of the *Quantity* of Good, and *Number* of Enjoyers." This led him to the important conclusion that "*that Action is best,* which accomplishes the *greatest Happiness* for the *greatest Numbers.*" Here is a precursor, by more than 50 years, to Jeremy Bentham's (1748–1832) utilitarian philosophy of "the greatest happiness for the greatest number."

In this new Hutchesonian human science, happiness is an important aspect of human behavior and existence. Recall the phrase in Jefferson's Declaration of Independence concerning the right to "life, liberty and the pursuit of happiness." The Hutcheson formula with its suggestion of calculability gave expression to beginnings of a numerical human science.

For the thinkers of the Enlightenment, this concept of the calculability of human happiness had specific implications, some of which are not obvious to us today. Happiness was a measurable quantity in two dimensions: both the amount of happiness within an individual and the sum of happiness within a group could be quantified. It was a concept central to the development of a new human science rooted in numbers.

HALES'S NUMERICAL PLANT AND ANIMAL SCIENCE

In the early eighteenth century, approximately a hundred years after William Harvey's pioneering work on the circulation of blood, Stephen Hales (1677–1761) offered a spectacular example of the application of numbers to the life sciences. A clergyman and amateur scientist, Hales is often called the founder of plant physiology. Hales believed that the phrase from the Book of Revelation that God had made the world by weight and measure implied that the way to understand the world of nature must be to weigh and to measure; that is, to base the science of nature on numerical data.

One quantity that Hales measured was the pressure of the sap in vines. In the early 1700s many people believed there was a circulation or an ebb and flow of sap in plants, a kind of analogue of the circulation of the blood in animals. Of course, there is no such circulation, because plants have no organ corresponding to the animal heart. However, the sap does rise.

Indeed, Hales made a major discovery about the movement of sap, as the result of a happy accident. An ardent gardener, he relates that he had pruned some vine stems in the springtime and was alarmed that the cut vines were bleeding or oozing sap. To

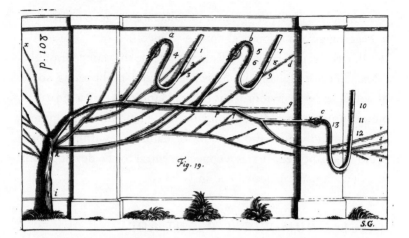

FIGURE 4.1 Stephen Hales's measurement of sap pressure (note that the vines are connected with three open manometers), from Hales's *Vegetable Staticks. Courtesy of the Botanical Library, Harvard University*

remedy this situation, he fastened little caps or covers made of pig's bladder around the cut tip of the vines. To his great surprise he observed that these caps or covers were expanding. Some kind of pressure was driving the sap upward in the springtime.[5]

Hales tells us that he at once changed conditions so as to convert his qualitative observations into numerical data. To this end he replaced the bladder caps by manometers (pressure gauges) (see figure 4.1), and thus became the first person in history to measure the phenomenon we know as root pressure, the pressure that drives the sap upward and outward to nourish the plant. All of us know of this flow of sap in the spring, since the sap of maple trees is collected from sugar-maple trees and then boiled down to produce maple syrup.

Hales made other contributions to plant physiology in his efforts to create quantitative science. His concern with numbers led him to be the first person to measure blood pressure in animals. Unfortunately, his method of determining blood pressure involved exsanguinating the animal. This procedure obviously could not be used for medical diagnosis in human beings. The poet Alexander Pope, Hales's neighbor, simply could not understand how so nice a man as Hales could perform experiments in which animals, as subjects for measurement, were allowed to bleed to death.

THOMAS JEFFERSON: A LIFE REGULATED BY NUMBERS

In the eighteenth century, the growing interest in using numbers to regulate and quantify ordinary experience began to find expression at the highest levels of political and social thought. Thomas Jefferson's intellectual world and his daily life were regulated by numbers to a degree that seems astonishing to a reader in the twenty-first century. Skilled in mathematics, he was, for instance, a master of the Newtonian calculus of fluxions and delighted in numbers and in calculation. Almost every aspect of his life was reduced to numerical observations and calculations. It is one great merit of Garry Wills's book on Jefferson and the Declaration of Independence to have described in detail Jefferson's commitment to numerical rules and quantitative observations. Wills traces Jefferson's concern with numbers to their possible sources, among them the writings of Sir William Petty, the prophet of a numerically based polity.

Jefferson kept detailed numerical records of his farming and gardening activities, of meteorological conditions, and of any other facet of daily life that could possibly be quantified. He exhibited his love of precision in a letter to Abigail Adams of 22 August 1813. He wrote that he had "ten and one-half grandchildren, and two and three-fourths great-grandchildren," adding "these fractions will ere long become units."[6]

He wanted to have almost every aspect of life reduced to numbers, and even his political thought was pervaded by this approach.

As a man of science of the post-Newtonian era, Jefferson knew that numerical evidence modifies abstract theory—a feature of exact science from the seventeenth century to the present. Jefferson was concerned for numbers not simply for their own sake, but as numerical evidence in the post-Newtonian manner. This idea appears again and again in Jefferson's writings. For example, he wrote how the "degrees"—that is, the actual numbers—"fix the laws of the animal and vegetable races, which may exist with us."[7]

At Monticello, Jefferson meticulously made and recorded daily observations ("with a good degree of exactness") of the meteorological conditions, the measurements of temperature, winds, rainfall, and barometric pressure. His data for a seven-year period from 1 January 1810 to 31 December 1816, he wrote, were so exact and continuous that they enabled him "to deduce the general results." There were in all "three thousand nine hundred and five" numerical observations.

"During the same seven years," he wrote, "there fell six hundred and twenty two rains, which gives eighty nine rains every year, or one for every four days." Furthermore, "the average of the water falling in the year being 47-1/2 inches, gives fifty three cents [i.e., hundredths] of an inch for each rain, or ninety three cents for a week."

One of the questions that interested Jefferson was whether a wheelbarrow with two wheels was more efficient than the customary one with a single wheel. Performing actual tests was the only way to find out. Jefferson's account of his measurements shows how he understood the importance of time and motion studies, two centuries before Frederick Taylor (1856–1915) did his pioneering research in the same field.[8]

Jefferson's extreme fascination with numbers also extended to politics, and could lead him to proposals that seem absurd to a reader in the twenty-first century. It is well known that Jefferson believed in periodic rebellions as a force for political renewal. As he wrote to James Madison in 1787, "I hold it, that a little rebellion, now and then, is a good thing, and as necessary in the political world as storms in the physical."[9] It is not as well known that he did not think we have a right to impose our ideas and regulations, and even our laws and contracts, on future generations. This is the sense of his many expressions of the need for future generations to free themselves from the chains of the past. The future, he held, should not be bound by what binds us in the present.

Such statements imply a need for future revolutions. Jefferson extended this proposed limitation to all pacts, constitutions, and agreements, believing that "the earth belongs in usufruct to the living."[10] (*The American Heritage Dictionary* defines "usufruct" as "the right to enjoy the profits and advantages of something belonging to another as long as the property is not damaged or altered in any way.") Being Jefferson, he quantified these notions and concluded that "every constitution then, and every law, naturally expires at the end of 19 years." In practice, the actual time would be only approximately 19 years. Jefferson rounded out this number to 20. On this subject Jefferson drew on the work of the French naturalist, Comte de Buffon (1707–1788).

In the spirit of charity, let us not explore the chaotic consequences that would follow had Jefferson's 19-year rule been put into practice. But this numerical fantasy is of interest in showing how Jefferson's thinking tended to be conditioned by his deep concern with numbers. Here is the seed of Jefferson's dictum that a society needs a revolution every 20 years.

Consider Jefferson's judgment on Shays' Rebellion, an uprising of economically depressed farmers in western Massachusetts that took place in 1786–87. "The late rebellion in Massachusetts," he wrote, "has given more alarm than I think it should have done."[11] He based this conclusion on simple numerical calculations. "Calculate that one rebellion in 13 states in the course of 11 years" is not very great, he wrote. This numerical ratio is the same as "one for each state in a century and a half." In fact, he wrote, "No country should be so long without one," adding that there is no "degree of power in the hands of government" that can "prevent insurrections." Writing from Paris, he took note that in France, despite "all it's [sic] despotism" and its "two or three hundred thousand men always in arms," he had been witness to "three insurrections in the three years" he had been there, "in every one of which greater numbers were engaged than in Massachusets [sic] and a great deal more blood was spilt."

Jefferson's concern with numbers and his arithmetical skill were of real importance for political action during his service as Secretary of State. One of the numerical problems addressed by Jefferson was an apparently simple one: how to assign to each state a fair number of representatives in Congress. This numerical exercise had to meet the provisions of the Constitution, which requires a review of apportionment every decade on the basis of the decennial census of the population. The Constitution, however, provides no details concerning the way an apportionment is to be made. A few rules were set forth. Here is the relevant text of the Constitution:

> Representatives and direct Taxes shall be apportioned among the several States which may be included within this Union, according to their respective Numbers, which shall be determined by adding to the whole Number of free Persons, including those bound to Service for a Term of Years, and excluding Indians not taxed, three fifths of all other Persons.

The actual Enumeration shall be made within three years after
the first meeting of the Congress of the United States, and
within every subsequent Term of ten Years, in such manner as
they shall by Law direct. The Number of Representatives
shall not exceed one for every thirty thousand, but each State
shall have at least one Representative; and until such enumeration shall be made, the state of New Hampshire shall be entitled to chuse [sic] three, Massachusetts eight, Rhode Island
and Providence Plantations one, Connecticut five, New York
six, New Jersey four, Pennsylvania eight, Delaware one,
Maryland six, Virginia ten, North Carolina five, South Carolina five, and Georgia three.[12]

There are certain conditions imposed by Article I. One
restriction is that the representation should not exceed one for
every 30,000 in the population. Women and children are to be
counted even though they had no voting rights. Furthermore, it
is provided that Negro slaves are to be included in the count but
only at the rate of three-fifths of a white person. A further condition is that each state must have at least one representative. The
same reckoning is to be applied to the obligation of the states to
support the expenses of the federal government.

The average reader of these conditions will assume that
apportionment is a matter of simple arithmetic. Most Americans
assume that the composition of the House of Representatives is
determined by a just and fair apportionment of representatives
according to the populations of the individual states. In fact,
however, very few people understand how that apportionment is
determined in actual practice. Determining the number of representatives for each state poses a mathematical as a well as a political problem, one that has been the cause of intense argument,
debate, and analysis during most of our country's history.

The problem of apportionment eventually has enlisted the

creative efforts of some of the country's most able mathematicians. Apparently simple and obvious solutions have proved to involve major political issues that have raised great passions of debate. The problem of dividing responsibility for the expenses of running the federal government does not present the same issues that beset apportionment, because a money obligation can be divided down to the last cent, but a representative cannot be divided.

Suppose that a given state has one-thirteenth of the total national population. Its share of the federal expenses can be easily and fairly reckoned at one-thirteenth of the whole.

In apportionment, however, there is no such easy division. Suppose that there are 120 seats in the House and that a certain state has one-thirteenth of the national population. Simple arithmetic gives this state one-thirteenth of the 120 representatives. Dividing 120 by 13 yields 9 3/13 or 9.231, a little more than nine representatives. Similar calculations for each state will yield an integral number of representatives plus a fractional or decimal part. Obviously, a state cannot have a "fractional" representative.

The simplest solution might seem to be to stick to the integral numbers and discard the fractional parts. This solution, however, would be unfair to a state entitled to, say, 1.987 representatives, reducing that state's representation from almost two to just one, a loss of nearly half of its just share. A loss of the fractional or decimal part would be less significant for a state entitled to 12.121 representatives, since its loss would be less than about one-tenth of its just share.

There are other possible solutions: one would be to give an extra seat to the state with the highest fraction, then to give the next extra seat to the state with the next highest fraction, continuing until the total number of representatives adds up to 120. An alternative would be to use a similar procedure, but start by giving an extra seat to the most populous state, rather than to the state with the highest fraction.

There are still other possibilities. Jefferson favored trying various divisors other than 120, discarding all fractions. Eventually, by trial and error, a divisor can be found that will yield a total number of 120 representatives.

Objections can be (and have been) raised to each of these methods, perhaps for favoring either the small or the large states or favoring the South versus the North. Clearly, there is no simple and obvious way to satisfy the requirements of the Constitution.

Congress first attempted to devise a system of apportionment in 1792 in legislation entitled "An Act for the apportionment of Representatives among the several States according to the first enumeration."[13] Thomas Jefferson analyzed this proposal for President George Washington and found such serious flaws that Washington vetoed the proposal. This was the first presidential veto in the history of our country, one of only two times that Washington exercised his veto power, showing how seriously Washington regarded the problems raised by Jefferson.

This incident provides a striking example of how blueprints for government may introduce technical problems that are beyond the scope of knowledge and forethought of the founders. It also displays the mathematical acumen of Thomas Jefferson, whose analysis of the proposed method provided both the grounds for Washington's veto and the basis for the system that was finally adopted.

Among the reasons that Jefferson objected to the first bill proposing an apportionment was that it contravened what he considered a basic principle of good government. The bill said nothing whatsoever about the actual method used in the proposed assignment. Only because Jefferson was skilled in arithmetic and the manipulation of numbers was he able to figure out how the apportioning had been done. (His analysis was confirmed when the method used by Congress was revealed by Alexander Hamilton to Washington in support of the bill.) Hence, as Jefferson quite correctly argued, when the next apportionment was needed, following the census of 1800, Congress would not know how it had been done the first time. Since the method had not been specified, the way was open to change the procedure at will the next time around. As Jefferson wrote, the bill "seems to have avoided establishing [the procedure] . . . into a rule, lest it might not suit on another occasion." Perhaps, he observed, "it may be found the next time more convenient to distribute them [the residuary representatives] *among the smaller states*; at another time *among the larger states*; at other times according to any other crotchet which ingenuity may invent, and the [political] combinations of the day give strength to carry." Jefferson wanted any law to contain an explicit method which "reduces the apportionment always to an arithmetical operation, about which no two men can ever possibly differ." This requirement of having a sound and unambiguous method was a fundamental characteristic of all good Enlightenment science, expressed simply and beautifully by Linnaeus, "Method [is] the soul of science."

Jefferson's application of arithmetic to a problem of statecraft is a splendid example of what the eighteenth century knew by William Petty's term "political arithmetic." This was a subject of great significance for Jefferson and many of his contemporaries in America and Europe, not the least of whom was Benjamin Franklin.

BENJAMIN FRANKLIN AND NUMBERS

Benjamin Franklin, like Jefferson, was fascinated by numbers, although he did not suffer from Jefferson's mania for numerations. He tells us in his autobiography that when he was about 16 years of age, he was "asham'd" of his "Ignorance in Figures," that is, arithmetic or reckoning, a subject which he "had twice fail'd in learning when at School." To remedy this situation, he obtained a book on arithmetic, which he went through by himself "with great ease." Practical applications were important. As a young tradesman in Philadelphia, he studied accounting and in his autobiography, Franklin advocated that young women be taught accounting. He had in mind that a woman might need to take over her husband's business in case he became ill or were to die.

His great success as a shopkeeper shows that he had mastered the subject. But sometimes Franklin used his skill with numbers simply to ease his boredom. As a young man, serving as clerk to the Pennsylvania Legislature, he found himself wearied by "sitting there to hear Debates in which as Clerk" he "could take no part" and which were often "unentertaining." So, to pass the time, he amused himself by performing numerical calculations for fun, producing numbers to fill out what are called magic squares and circles.

A magic square is a square array of numbers, in which the sum of the numbers in each column and in each row is the same and is equal to the sum of the numbers in each of the diagonals. A famous magic square appears in Albrecht Dürer's sixteenth-century engraving of *Melencolia I*. Here (see figure 4.2) the sum of the numbers in each row (e.g., 16 + 3 + 2 + 13) is 34. The sum of the numbers in each column (e.g., 16 + 5 + 9 + 4) is also 34, as is the sum of the numbers in each of the two diagonals (16 + 10 + 7 + 1 or 13 + 11 + 6 + 4). (See figure 4.3.)

Later on, Franklin wrote of how he had "acquired such a knack at it," that he could "fill the cells of any magic square, of reasonable size, with a series of numbers as fast as I could write them."[14] Finding the traditional magic squares to be too "common and easy," he devised more complex forms of magic squares

FIGURE 4.2 Albrecht Dürer, *Melencolia I*, 1514. *Courtesy of the Fogg Art Museum, Harvard University Art Museums, Gift of William Gray from the collection of Francis Calley Gray*

"with a variety of properties." He proudly described one of these as "the most magically magical of any magic square ever made by any magician." Indeed, not content with his own form of magic square, Franklin went on to invent a magic circle. So proud was

16	3	2	13
5	10	11	8
9	6	7	12
4	15	14	1

FIGURE 4.3 The magic square from Dürer's *Melencolia I.*

he of his magic squares and circles that Franklin included one of each in the revised and expanded 1769 edition of his book on electricity.[15]

Of course Franklin did not invent the concept of a magic square, nor do we know exactly how he learned about these arithmetic curiosities. However, some years later he described his encounters with two books that mentioned magic squares. In a letter to his London patron, Peter Collinson, Franklin explained how he had encountered a book on the subject in the library of James Logan, a Philadelphian who owned the most extensive collection of scientific books in Colonial America. Logan had shown him a "remarkable" book on magic squares. It was, Franklin recalled, "a folio French book, filled with magic squares, wrote, if I forget not, by one M. Frenicle."[16] Logan also showed Franklin "an old arithmetical book, in quarto, wrote, I think, by one Stifelius, which contained a square of 16." Franklin recalled how "that evening," at home, he produced a magic square of 16 with unusual properties, which he later reproduced in his book on electricity. This is the same Stifel (or Stifelius) whom we have seen proving Pope Leo X to be the anti-Christ because the letters in his name could add up to 666, the "number of the Beast."

Franklin's appreciation of the role of numbers in statecraft was manifested early in his career. In 1729 (when he was 25 years old) he wrote a pamphlet on the need for paper currency. Franklin based his ideas on currency on Petty's "political arithmetic," and later, stimulated by Petty's ideas, he developed the concept of a labor theory of value.

The labor theory of value, in Franklin's words, holds that "Trade in general being nothing else but the exchange of Labour for Labour, the value of all things is . . . most justly valued by Labour."[17] That is, the value of any commodity depends on the labor needed to produce it. This theory eliminates such factors as capital investment and market scarcity.

The full title of Franklin's tract is *A Modest Enquiry into the Nature and Necessity of a Paper-Currency.* In it, Franklin aimed to show that value need not be measured only in terms of the precious metals, gold and silver. These have long been admired, he wrote, for their "Fineness, Beauty, and Scarcity." But even though, commonly, all things are valued by silver equivalent, silver itself is "of no certain permanent Value, being worth more or less according to its Scarcity or Plenty." Therefore, "It seems requisite to fix upon something else," something "more proper to be made a *Measure of Values,* and this I take to be *Labour.*"[18]

To make this concept clear, Franklin gave an example, actually a "close paraphrase" of two passages in Petty's Treatise of Taxes (1662).[19] In this example, it is supposed that two men are at work, one raising corn, the other "digging and refining" silver; at the end of a year (or other suitable interval of time), one man will have grown 20 bushels of corn while the other will have produced 20 ounces of silver. Therefore an ounce of silver is worth the labor of growing a bushel of corn. The actual worth of silver may decline, as happened with the discovery of the mineral riches of the New World, but this does not directly affect the rate of labor expended in producing a bushel of corn.

Franklin's concern with numbers as the exact expression of

exact information is evident in the pages of the *Pennsylvania Gazette*, the weekly newspaper that Franklin edited and published in his late twenties and early thirties. On this topic I draw on information assembled by the dean of Franklin scholars, J. A. Leo LeMay, for his forthcoming magisterial biography of Franklin. LeMay found that the pages of the *Gazette* regularly displayed numerical data. Franklin, as a man of his times, read books with quantitative data, the eighteenth-century works that advanced the subject of political arithmetic. Franklin also drew on compilations of data concerning trade and shipping.

Thus he informed his readers about the work of Joshua Gee, who had had a certain success in London for his books containing numerical data on trade and population. On 5 January 1731, Franklin published in the *Gazette* what LeMay describes as "information concerning all ships that had entered or cleared out of the major colonial ports" except for Southern ports for which there was apparently no information readily available. This shipping news was not a space filler, but a subject of real importance, and on the front page Franklin called his readers' attention to these data. "In this paper," he announced, "we exhibit an Account for one Year, of all the Vessels entered and cleared, from and to what places, in the Ports of Philadelphia, Amboy, New-York, Rhode-Island, Boston, Salem and New Hampshire." These data, he explained, would enable the "ingenious Reader" to "Make some Judgment of the different Share each Colony possesses of the several Branches of Trade." To provide a complete picture of the situation, Franklin printed an extract from Gee's *Trade and Navigation of Great Britain Considered*, dealing with the trade between Britain and the American colonies. A supplement appeared on 12 April 1731 with trade data for Barbados. Franklin's regular publication of such data concerning trade was a popular feature of the *Pennsylvania Gazette*.

According to LeMay, Franklin's interest in the numerical

aspects of trade "paralleled his interest in demography." Thus, on 20 November 1729, he began to publish data on the number of burials in Philadelphia, first week by week, and then totaled for the year. So that his readers might grasp the significance of these numerical data, he published comparative numbers for Boston, Berlin, Amsterdam, and London.

On 6 August 1731, Franklin published data on the inhabitants of Breslau, noting that one twenty-ninth of the population died each year. The data for Boston, he found, showed that "not above a 40th Part of the People of that Place die yearly, as a medium." In evaluating the use of such numerical information, we should keep in mind the warning of James Cassedy, a historian of early American medical statistics. Cassedy found that mortality data were regularly skewed to an optimistic view of the healthy conditions of life in Colonial American cities.[20]

In the present context, these signs of Franklin's early concern with demographic data are worth noting because, as we shall see shortly, Franklin's continuing interest in demography led to his devising a law of population increase and eventually provided a basis for a policy that he advocated for Britain and the Colonies.

FRANKLIN AND MALTHUS

Franklin's interest in numbers appeared in his contributions to the new science of demography, the numerical study of populations. One major figure in this area was Thomas Malthus (1766–1834), an English pioneer who set forth the basic rule of this new science at the end of the eighteenth century. Malthus's rule states that if there were no barriers to the growth of human populations, there would be a geometric increase in the size of any population. Another way of saying this is that an unchecked population will double in size every so many years.

This rate of growth, also called *exponential*, produces an

increase at an unbelievably large rate. A story is told of a native of India who had performed an important service for his king. The king offered him any reward he wished. He replied that he was a humble man and would ask only for a grain of rice to be placed on the first square of a chessboard, then two grains on the second square, then four on the third, and so on until the last square. The king was sorry that his loyal subject had asked for such an apparently small gift. But, in fact, this program—if actually carried out—would have required more rice than was available in all of India. This example shows the enormous rate of increase in a geometric expansion.

Malthus also is credited with a second and related law—that the maximum possible increase in the food supply is in a simple arithmetic ratio. That is, in two years you can double the food supply, in three years triple it, and so on. This is a vastly slower rate of increase than the geometric or exponential growth rate mentioned above.

Darwin extended these laws of Malthus from human populations to all animal and vegetable populations. Then, in a flash of insight, Darwin recognized that all of the individuals of such geometrically increasing populations couldn't survive. And so Darwin was led to the doctrine of "natural selection" and the theory of evolution by "natural selection." Clearly, the law of population growth has been of enormous importance for science.

In the second edition of his book on population, published in 1803, Malthus declared that the rule of population doubling had been stated a half-century or so earlier by Benjamin Franklin in his writings on demography. It is difficult to be sure who first stated this important law, but many thinkers of the late eighteenth century associated this law with Benjamin Franklin.

Franklin wrote two works on the new science of demography. The first was a pamphlet entitled "Observations Concerning the Increase of Mankind."[21] Written in 1751, this essay was first published in Boston in 1755 and reprinted soon after in London. Later, with various revisions, it appeared in a dozen or so different publications, becoming one of Franklin's most often reprinted essays.

The "Observations" was written in response to the British Iron Act of 1750, which restricted the manufacture of iron in the American colonies. Franklin's argument employed political arithmetic, based on the idea that policy issues should be determined by a statistical analysis of numerical data.

Franklin's important contribution to demography centered on his clearly stated rule that, under the American conditions of relatively unchecked growth, the population would double every 20 or 25 years. The major part of this pamphlet is devoted to numerical demographic data concerning population—births, deaths, and marriages—from all over the world. Franklin's law of population growth is thus firmly rooted in numbers and it leads to the conclusion that land is needed for this expanding population. Additionally, Franklin's law leads to an important principle of policy: the recognition that there will be a time when "the greatest Number of *Englishmen* will be on this Side of the Water."

Accordingly, Franklin concluded that an expansionist policy in North America was a necessity and that British America was destined to become the most populous and important part of the British empire. After 1751, the "increase of mankind" became the very core of Franklin's faith in the inescapable growth of American power, either within the framework of the British empire or without and even against it. Clearly Franklin's theoretical or scientific work on demography led to practical political considerations of national policy.

A second publication in which Franklin expounded his ideas on population is the so-called Canada pamphlet, actually entitled "The Interest of Great Britain Considered, With Regard to her

Colonies, And the Acquisitions of Canada and Guadaloupe."[22] This work was composed and published in 1760, while Franklin was colonial agent in London. The occasion was the impending favorable conclusion of the Seven Years' War, known by Americans as the French and Indian War. As victors, the British would be able to annex either Canada or Guadeloupe.

In his booklet, Franklin included the earlier "Observations," providing a very strong argument for the acquisition of Canada. The increase in population, of which he had written almost 10 years earlier, would require new regions into which the population could spread, thus peopling a greater part of North America with British colonials.

Franklin argued that with protection from French foes and their Indian allies, and with the availability of cheap land, there would be a natural increase of population. The consequence would be an ever-expanding market for British manufactured goods.

Suppose now, Franklin argued, that Britain chose Guadeloupe and that there was no room for natural expansion in British North America. In this case, the colonists would be "confined within the mountains." Under these conditions, the natural increase of population would cause the population density to increase until it became as great as that of Britain. The cost of land would rise and wages would fall. The industries—extractive, plus agriculture and hunting—would, under these circumstances, no longer be as profitable as before and the colonists would be forced to turn to manufacturing.

Under these conditions, Americans would become producers rather than consumers. Hence Americans would depend less and less on the mother country. It was clearly to the advantage of Britain to annex Canada. Historians generally agree that Franklin's numerically based argument was a major influence on Britain's decision to acquire Canada rather than Guadeloupe.

FRANKLIN ON NUMBERS AND SMALLPOX

A spectacular example of Franklin's use of numbers is provided by his public advocacy of inoculation. Franklin was particularly sensitive about this controversial practice because, when his son Francis died of smallpox, rumors circulated that the boy had contracted the disease through inoculation. To counter such rumors, Franklin published a notice in the *Pennsylvania Gazette,* announcing that his son had contracted smallpox in the normal or natural way.[23]

Smallpox has been essentially eliminated from the world disease picture, but in Franklin's day and earlier, smallpox epidemics would sweep through Europe and America and there was real fear of this cause of death. At that time inoculation was the only known method of prevention.

Inoculation differs in an important way from the later practice of vaccination. The term *vaccination* has its root in the word "vaccinia," meaning cowpox, a rather mild disease related to smallpox. In vaccination, the patient is exposed to the viral material of cowpox lesions, causing him or her to become immune to cowpox; this immunity also confers on the patient an immunity to the related and more deadly smallpox. Edward Jenner (1749–1823), a British physician, discovered this method of preventing smallpox in the late eighteenth century.

In inoculation, on the other hand, the patient is exposed to the viral material of smallpox itself, taken from the lesions of a smallpox patient. This method usually gives the patient a relatively mild case of smallpox, which does not kill the patient but leaves him or her immune to the disease.

Inoculation had its own dangers, and some patients who had been inoculated died of the infection. A well-known case was that of Jonathan Edwards, the celebrated colonial intellectual figure and preacher in Northampton, Massachusetts. Appointed president of the college now known as Princeton, where there was a raging epidemic of smallpox, Edwards had himself and his family inoculated before moving to New Jersey, only to die from the inoculation before taking office. But inoculation was not generally fatal. The Dutch physician Jan Ingenhousz, a friend of Franklin's, was employed for many decades as physician to the royal family of Austria. His chief job was to perform inoculations, and his record was perfect: not a single death from inoculation!

In America feelings ran high on whether to inoculate one's family. One way for a person to decide whether to have his children inoculated was to appeal to numbers. But what were the actual numbers? What was the probability of getting smallpox in an epidemic, and how did these numbers compare with his or her chances of death? And how did these numbers compare with the probability of death resulting from inoculation? These topics were central to Franklin's short essay on smallpox and inoculation, published in his newspaper in 1736.

Franklin began by referring to a "current Report, that my Son Francis, who died lately of the Small Pox, had it by Inoculation."[24] Franklin was concerned that some people might be deterred "from having that Operation perform'd on their children" on the basis of "that Report (join'd with others of the like kind, and perhaps equally groundless)." Accordingly, Franklin did "hereby sincerely declare," that his son had not been inoculated, but had "receiv'd the Distemper in the common Way of Infection." Franklin added that he supposed the report of his son's death must have arisen from "its being my known Opinion, that Inoculation was a safe and beneficial Practice; and from my having said among my Acquaintance, that I intended to have my Child inoculated, as soon as he should have recovered sufficient Strength from a Flux with which he had been long afflicted."

In Part Three of his autobiography he reiterated these sentiments:

In 1736 I lost one of my Sons a fine Boy of 4 Years old, by the
Small Pox taken in the common way. I long regretted bitterly
& still regret that I had not given it to him by Inoculation;
This I mention for the Sake of Parents, who omit that Operation on the Supposition that they should never forgive themselves if a Child died under it; my Example showing that the
Regret may be the same either way, and that therefore the
safer should be chosen.

Several decades later, when Franklin was in London as agent
for several American colonies, he was active in promoting inoculation. He joined forces in this endeavor with William Heberden,
a prominent London physician, who—like Franklin—was a fellow of the Royal Society. The two of them produced a pamphlet,
for distribution in the American colonies, which set forth the
way to perform inoculations. Franklin wrote an accompanying
essay on the history of the practice and gave numerical evidence
to show that this practice was relatively safe. Franklin was very
proud of this essay and had it reprinted in 1759 as a separate
work. The pamphlet was entitled "Some Account of the Success
of Inoculation for the Smallpox in England and America."[25]
Franklin's essay begins with a history of the practice of inoculation in New England. At the outset, Franklin showed his deep
understanding of the use of numbers in any controversy. He
explains that the data on this topic tended to be unreliable. The
"practice of Inoculation always divided people into parties," he
wrote, because some people would be "contending warmly for
it" with "others as strongly against it." Those opposed to inoculation would assert that "the advantages pretended were imagi-
nary," that the "Surgeons, from views of interest, conceal'd or
diminish'd the true number of deaths occasion'd by Inoculation,
and magnify'd the number of those who died of the Small-pox in
the common way." Accordingly, the job of reporting on the

number of deaths by inoculation and ordinary smallpox was turned over to town constables who had to submit their numbers under oath.

Our interest in Franklin's pamphlet centers on the use of numbers in a policy debate and on Franklin's recognition that in evaluating any medical practice, the test must be based on numbers. Those of us old enough to remember the introduction of the Salk vaccine for polio will recall that the preliminary tests were numerical, based on a massive nationwide statistical analysis.

Franklin's goal was simple and straightforward. He wanted to give anxious parents evidence that it was safe to have their children inoculated. The data he assembled were most impressive. For example, Franklin presented data he obtained from "Dr. Archer, physician to the Small-pox hospital here." During the period from its founding to 31 December 1758 there had been given a total of 1,601 inoculations. From these, only six recipients had died. During these same years the number of "Patients who had the Small-pox in the common way" were 3,856, of whom 1,002 had died. In other words, the risk of death from the mild case of smallpox produced by inoculation was minuscule (about 3 out of 800), whereas the chance of death from smallpox taken "in the common way" was rather high (about one out of four).

Data from the Foundling Hospital were even more impressive, Franklin noted. In this hospital, the practice was that "all the children admitted, that have not had the Small-pox, are inoculated at the age of five years." There had been 338 children inoculated since this practice had been put into effect. Of this number, Franklin reported, only two had died. One of this pair had, in fact, not died from the inoculation, but had been a victim of a "worm fever."

"On the whole," Franklin concluded, "if the chance were only as *two* to *one* in favour of the practice among children, would it not be sufficient to induce a tender parent to lay hold of the advantage?" But, "when it is so much greater, as it appears to be by these accounts (in some even as *thirty* to *one*)," then no parent would any longer "refuse to accept and thankfully use a discovery God in his mercy has been pleased to bless mankind with."

Franklin was not always in dead earnest when writing about numbers. In 1755, he penned a letter about marriage to Catherine Ray, a young woman to whom he was sending some "fatherly Advice."[26] This letter shows his delight in making puns, comparing the ingredients of a happy marriage to the four basic operations of arithmetic.

"You must practise *Addition* to your Husband's Estate," Franklin wrote, "by Industry and Frugality" and "*Subtraction* of all unnecessary Expences." As to "*Multiplication,*" he wrote, "I would gladly have taught you that myself," but "you thought it was time enough, and wou'dn't learn"; now it would be the husband who "will soon make you a Mistress of it." As to "*Division,*" however, "I say with Brother Paul, *Let there be no Divisions among ye.*"

Franklin and Jefferson brought their fascination with numbers into the creation of the new republic, which emerged as a political entity just as their importance as instruments of policy was becoming clear even to people who were not geniuses.

5 NEW USES FOR NUMBERS

NUMBERS AND MEASURES

Historians note two features of the use of numbers in the late eighteenth and early nineteenth centuries. First, the messianic "quantifying spirit" of the age[1] pushed numerical considerations into new domains. For example, numerical methods were introduced into areas of medicine that had previously seemed nonnumerical, such as the treatment of insanity. Second, measurements at this time attained a new high level of accuracy, making this an "age of precision."[2]

Both features are evident in the field of statecraft, in the actual counts that replaced estimates of national populations and in attempts to ascertain the actual production and consumption of food.

The quantifying spirit of the age particularly appears in the introduction of national censuses. In 1789 the new American Constitution called for a decennial census. The first federal census was taken in 1790. National censuses were taken in Sweden in 1792 (replacing earlier estimates of the size of the population), in Holland in 1795, in Norway and Denmark in 1797, and in England in 1801.[3] The American historian of statistics, Helen Walker, surmised that fear of divine punishment for a sin like King David's was removed by the obvious flourishing of the United States after the census of 1790.

In France, an attempt had been made in 1697 to estimate the size of the population. Instructions were sent out to the 32 "intendants" or regional governors appointed by Louis XIV, ordering them to "report the number of towns, villages, hamlets, and inhabitants within their jurisdictions." The intendants, however, "had neither the means nor the inclination" for the collection of this kind of demographic data. They returned a "medley of back numbers" taken from tax records and "counts of hearths" rather than making a true head count. Nine of the 32 intendants made no report at all.[4]

In 1800 the French government established a Bureau de Statistique Générale and in that same year issued a decree that there be a census of all the inhabitants of France. This census has been described as a "great Failure." The actual count was made under the direction of Lucien Bonaparte, Napoleon's brother, then Minister of the Interior. Lucien apparently believed that the census could be completed in a few months, but in fact the head count and tabulation of data required two years.[5]

These census figures seem to us to be innocent data, but in the eighteenth century the knowledge of such numbers could be of strategic importance in estimating a nation's potential military strength. Such security considerations apparently induced Sweden during the 1760s to keep its census data secret under the control of an Office of Tables.

In an essay on the quantifying spirit in the thought of the late eighteenth century, John Heilbron has called attention to a "newly discovered effective emphasis on precision."[6] This precision appears as an increase in the number and variety of concepts quantified, and as an increase in the accuracy of measurements. Precision and accuracy of measurement imply that a concept has been reduced to numbers. Not surprisingly, some leading thinkers of the time envisioned that all of human thought might be expressed mathematically.

Two examples display the enormous increase in precision at this period. From the time of Tycho Brahe (1546–1601) in the late sixteenth century to the age of John Flamsteed (1646–1719), accuracy in astronomical observations increased from about 1 minute of arc to 20 seconds, a factor of 3. In contrast, during the eighteenth century accuracy improved by a factor of 200. This same increase in precision can be seen in the measurement of time. The first pendulum clocks, built in the late seventeenth century, were good to some 10 seconds a day, but by 1800 chronometers were good to one-fifth of a second per day.

A CONCERN WITH NUMBERS IN FRANCE: LAVOISIER'S ESSAY ON POLITICAL ARITHMETIC

Antoine-Laurent Lavoisier (1743–1794), one of France's most distinguished scientists, is generally known as a primary founder of the modern science of chemistry. He established the division of substances into elements, compounds, and mixtures, a distinction that we still use.[7] He spearheaded the creation of a new rational nomenclature of varieties of matter according to their chemical composition. For example, the traditional pre-Lavoisier name for a certain substance was "blue vitriol." In the new system, the name was based on the chemical composition. Blue vitriol turns out to be a compound of copper, sulfur, and oxygen—a fact expressed in the name "copper sulfate." If the substance had been a compound of copper and sulfur without oxygen, the name would have been "copper sulfide." So great were the innovations introduced into chemistry by Lavoisier and his associates that the change is generally known as the Chemical Revolution.

One of the most fundamental principles of Lavoisier's chemistry was the use of numbers, notably in relation to what we often call today the principle of conservation of mass: in a chemical reaction, mass is neither created nor destroyed. In other words, in every chemical reaction, the weight of all the reacting substances is always equal to the weight of all the product substances. This principle implies that an experimenter must not only keep account of all the reacting solids and liquids, but also the gases—that is, all of the products. In particular, Lavoisier's rule directed attention to the reacting and product gases. This rule led to quantitative experiments. Lavoisier was not the first person to use numbers in chemistry but he was a pioneer in using such numerical measurements as the basis of his system of chemistry.

Lavoisier's law implies the indestructibility of matter in chemical reactions: matter is neither created nor destroyed. In Lavoisier's own words: "Nothing is created in the operations either of art or of nature, and it can be taken as an axiom that in every operation an equal quantity of matter exists both before and after the operation."[8]

When Lavoisier first announced this law, chemists generally believed in something called "phlogiston" which supposedly entered into chemical reactions (such as combustion) but had no weight. It was a radical step, therefore, for Lavoisier to base a system of chemistry on a balance of weights and to maintain that chemistry is not concerned with weightless "substances." In a very real sense, this was indeed a chemical revolution.

Lavoisier was an enthusiastic supporter of the French Revolution in its early and moderate phase. In a letter to Benjamin Franklin in 1790 he referred to the new chemistry as a revolution, saying he would consider it "well advanced and even completely accomplished if you range yourself with us." Lavoisier continued, "After having brought you up to date on what is going on in chemistry, it would be well to speak to you about our political revolution. We regard it as done and without any possibility of return to the old order." Clearly the two revolutions were linked in his mind. However, during the later stages of the Revolution, Lavoisier was arrested because under the former government he had been a staff member of the "ferme générale" or tax collection agency, and on 8 May 1794 he was decapitated on the guillotine.

Before the Terror, however, Lavoisier worked in many areas for the new government. Because of his concern with numbers and precision, Lavoisier was made a member of the commission to establish the metric system (adopted in 1799).

One of Lavoisier's assignments led him directly into numerical statecraft: making a study of "the territorial wealth of France," a census of the land actually under cultivation for farm production. His report was published by order of the National Assembly in 1791 as the lead chapter in a work enti- tled *Collection de divers ouvrages d'arithmétique politique par Lavoisier, Delagrange et autres* (Paris: An IV [1791]). The sec- ond work in this collection was written by the eminent mathe- matician Joseph-Louis Lagrange (1736–1813), a close friend of Lavoisier's.

The new French Republic needed Lavoisier's survey because taxation was based on property, on land actually under cultiva- tion or in use for raising livestock. His method of estimating the amount of land under cultivation was ingenious.[9] He assembled data on the annual consumption of food and wine in urban and in rural households and then computed how much land would have to be under cultivation to produce this quantity of food and drink. He also insisted that this method of computing the area of land under cultivation was possible because there were at that time no sizable imports or exports of food. Yet, the lack of reli- able figures for the total population of France significantly flawed his method. Jean-Claude Perrot, a scholar who has made a detailed study of Lavoisier's work, points out that Lavoisier underestimated the size of the population of France by 2.5 million.[10]

Lavoisier estimated the total population of France to be some 25 million, of whom he supposed 8 million lived in towns. He recorded how many people were engaged in various activities. For example, he found that 2.5 million people were engaged in viniculture. He reported that in France the annual consumption of grains (wheat, rye, barley), including both seed and comestible grains, came to 1,400 "livres pesant."

To estimate how many horses and oxen were employed in agriculture, Lavoisier started with the amount of grain consumed by the French population. He reckoned the area of cultivated land of average productivity needed to produce that amount of grain, the number of plows needed to cultivate that area, the number of oxen or horses required to draw a plow through soil of average density, the relative efficiency of plows drawn by oxen and horses, and the proportion of oxen and horses employed in plowing as an average over the country. He calculated the area of land plowed by each kind of draft animal, and finally calculated an estimate of their actual numbers.[11]

He recognized that to get the total number of horses in France he would also have to estimate the number of horses not used in agriculture, such as those in cities and in haulage. At this point in his analysis, he admitted that these numbers were obviously "fort hypothétique" (extremely hypothetical). He concluded with estimates of the total number of cattle, sheep, and pigs. One of his remarkable conclusions was that only one-third of the arable land of France was actually under cultivation.

Lavoisier ended by calling for a permanent office of facts and figures, a statistical bureau, to keep regular records of agriculture, commerce, and the size of the population of France.[12]

According to Lavoisier, only in one country—France—could such a bureau be established and maintained. This project, he concluded, depended on "the will of the Assemblée Nationale." He added that with the establishment of such a bureau and the collection of factual data, the "science of political economy" would cease to exist because all problems would be solved with no disagreements whatsoever.[13]

SIR JOHN SINCLAIR'S CENSUS OF SCOTLAND

Sir John Sinclair's (1754–1835) statistical account of Scotland affords further evidence of the growing concern with national and social numbers at the end of the eighteenth century. In the words of Sinclair's biographer, this was a time marked by a tendency of "the government to go around counting things."[14] This was in part "the result of a developing social conscience in the political part of the nation." But it also came from "an immediate need to check up on food and manpower supplies in a war that was revealing shortages of both." A number of individuals began to search for exact—that is, numerical—information concerning social conditions. For example, in 1794 Jeremy Bentham was "drawing up a table of the property and population of the country." Recall that the first census in Britain took place in 1801.

Sinclair, a wealthy Scot and a member of Parliament, was particularly interested in agriculture and had a passion for numerical information. His census was planned to be more than a mere head count. Such a count had been made a half-century earlier by Dr. Alexander Webster (1707–1784),[15] who obtained his numerical data from parish ministers, the same source of information to be used by Sinclair half a century later. Sinclair's method was to send out a questionnaire to every minister in Scotland. He had the backing of the Church of Scotland and used both financial pressure and cajolery to get these informants to return the forms. He used his own funds to staff a secretariat to process the forms as they came in to the central office. Obviously, some ministers were more conscientious than others, and so the standard of reporting was rather uneven. One obvious fault of this system was that the head count did not include Roman Catholics, freethinkers, Jews, and those who simply were not churchgoers. But in Scotland at this time these would not represent large percentages of the total population.

The publication of these ministerial reports came to 21 volumes published from 1791 to 1799, of which volume 20 contained a description of Sinclair's method and a summary of results. These reports of the ministers were not edited or printed in regional groups. Rather, Sinclair put them into print as they came in to the central office.

A mere description in words cannot give the reader any sense of the monumental scale of this undertaking and the wealth of social and economic data Sinclair's volumes contain. A single example will give some sense of the information in Sinclair's census. A Mr. Smill of Dornock in Dunfriesshire reported concerning the annual finances of a "common labourer" with a wife and four children. Table 5.1 gives the family's living expenses for a year. The annual earnings of this man came to 14 pounds, 8 shillings, less than his living costs. This difference was made up by the wife's earnings, working as an agricultural laborer during harvest and haying seasons, and spinning wool during the winter and spring. Whatever their faults, these data are of such value to social and economic historians that the contents of these volumes are currently being edited for publication reorganized by locality and with detailed indexes.

One final aspect of these volumes should be noted. Sinclair boldly titled these volumes *Statistical Account of Scotland*. Here I use the adverb "boldly" because at that time the adjective "statistical" had not as yet come into general use in the modern sense. As Sinclair wrote:

> Many people were at first surprised, at my using the new words, *Statistics* and *Statistical*, as it was supposed, that some term in our own language, might have expressed the same meaning. But, in the course of a very extensive tour, through the northern parts of Europe, which I happened to take in 1786, I found, that in Germany they were engaged in a species of political inquiry, to which they had given the name of *Statistics*; and though I apply a different idea to that word, for by Statistical is meant in Germany, an inquiry for the purpose of ascertaining the political strength of a country, or questions respecting *matters of state*, whereas, the idea I annex to the term, is an inquiry into the state of a country, *for the purpose of ascertaining the* quantum *of happiness enjoyed by its inhabitants, and the means of its future improvement;* yet, as I thought that a new word might attract more public attention, I resolved on adopting it, and I hope that it is now completely naturalized and incorporated with our language.

At the dawn of the nineteenth century the word "statistics" had two unrelated meanings. In Germany, the term "Statistik" referred to a nonmathematical form of statecraft, the collection of information of the sort that used to be taught as political geography. At the same time, numerical statecraft was being pursued in a form which agrees with the word "statistics" as we understand it today. Toward the end of the nineteenth century, this confusing situation was described by the words "two roots," to indicate that there were two unrelated forms of study which are related to the emergence of the modern concept of statistics. However, someone who pursued statistics was still called a "statist" as late as 1878 when a French calculating machine was brought to the attention of English statisticians.[16]

TABLE 5.1 ANNUAL EXPENSES OF A "COMMON LABOURER" OF DUNFRIESSHIRE

	£ (1 pound = 20 shillings)	s (1 shilling = 12 pence)	d (1 penny)
House-rent, with a small garden or kailyard	1	0	0
Peats or fuel	0	6	0
A working jacket and breeches, about	0	5	0
Two shirts, 6s. a pair of clogs, 3s. 2 pair of stockings, 2s.	0	11	0
A hat, 1s. a handkerchief, 1s. 6d.	0	2	6
A petticoat, bedgown, shift and caps for the wife	0	9	0
A pair of stockings, 1s. clogs, 2s. 6d. apron, 1s. 6d. napkin, 1s. 6d. for ditto	0	6	6

	£	s	d
A shirt 2s. clogs, 2s. stockings, 2s. for each of the four children	1	0	0
Other clothes for the children, about 4s. each	0	16	0
School wages, etc. for the four children	0	10	0
Two stone of oat meal, per week at 20d. per stone	8	13	4
Milk, 9d. per week; butter, 3d. per ditto	2	12	0
Salt, candle, thread, soap, sugar and tea	0	13	0
The tear and wear of the man and wife's Sunday clothes	0	10	0
Total outlays	17 £	14 s.	4 d.

From Sir John Sinclair, ed., The Statistical Account of Scotland, 1791–1799 *(Edinburgh: W. Creech, 1799; reprint East Ardsley, England: E. P. Publishing, 1977)*

PINEL'S MEDICAL NUMBERS

The growing concern with numbers at the opening of the nineteenth century is apparent in the changes in the treatment of the insane in France. This development is all the more remarkable in that, until the development of pharmaceutical treatments for biologically based mental illness, this area of medicine had not been considered to be a science using quantitative evidence. The writings of Sigmund Freud (1856–1939), for example, do not contain references to numerical evidence.

Philippe Pinel (1749–1826) is honored in the history of medicine because he radically altered the ways in which people who were "insane" were treated. At the time of the French Revolution, men and women who were suffering from mental disorders were apt to be chained in dungeons and generally treated as if they were savage beasts. In 1793, when Pinel received a post at l'hôpital de Bicêtre, an institution in Paris which housed insane inmates, he found that one male patient had been restrained by chains for 36 years![17] Pinel introduced a new point of view, arguing that such patients were "sick" and should be treated with kindness. A famous painting by Tony Robert-Fleury shows Pinel freeing a patient from shackles. The painting is entitled *Pinel délivrant les aliénés* with the masculine "aliénés" even though it is a young woman who is being freed.[18] Pinel believed that all individuals should be protected against forcible restraint.[19] It has been suggested that Pinel's treatment of the insane was in harmony with the ideals of the French Revolution, which made it a duty to protect all individuals, even the insane, against chains.

Our concern, however, is not with Pinel's role as a founder of psychiatry, nor as "the greatest hospital superintendent of all times."[20] Rather, our focus is his "passion for statistical information." Pinel described his method as the new "calculus of probability," but, as one scholar notes, his research involved "little more than a judicious use of arithmetic," so that "calculation of proportions" is a better description of the procedure. Pinel kept careful daily records of his patients and he made from them numerical tabulations "from which comparisons could be made between subgroups. . . . [He was] one of the great disease classifiers."[21]

Pinel was, above all, an empiricist, and as such he did not bow down to established authorities. He wanted to avoid getting "lost in vague arguments about objects inaccessible to human understanding." Pinel's major work was his *Traité médico-philosophique sur l'aliénation mentale* (Paris: Brosson, 1809). In the introduction he stated his philosophy as follows: "[A] wise man has something better to do than to boast of his cures, namely to be always self-critical."[22]

In his research Pinel faced two problems: how to classify the mental disorders he encountered in patients and how to evaluate the success or failure of his therapeutic measures. He applied numerical methods to both.

In his analysis of Pinel's work, the late William Coleman observed that Pinel argued that his numerical method "had already served with good effect for the study of objects of social life." Numerical methods would "kill ungrounded speculation" and allow medicine ("the healing art") to escape from "blind empiricism" and to become a "true science."[23] As an example of his "numerical method," Pinel reported on patients he had treated over a four-year period. He divided his therapeutic outcomes into two categories: cured and discharged, or not cured.

Pinel did not succeed in establishing the treatment of mental disorders on a sound numerical basis—the task was too complex and too little understood for one person to achieve this end. But his example does show us how the goal of a numerically based science had permeated even the least mathematical part of medicine, treatment of mental illness.

One of Pinel's American students was a Boston physician, Dr. George Parkman (1799–1849). Dr. Parkman later gained notoriety in the annals of crime as being the only person to have been murdered by a Harvard professor. Dr. John White Webster, a professor of chemistry, cut up Parkman's corpse and fed the pieces into a furnace in his chemistry laboratory, hoping to destroy the evidence against him. But Dr. Parkman's teeth, found in Dr. Webster's furnace, were enough to convict Webster.

LOUIS AND THE NUMERICAL METHOD

Historians and statisticians generally agree that Pierre Charles Alexandre Louis (1787–1872) should be honored as a primary author of the "numerical method." By "numerical method" he meant that the stages of disease and therapeutic outcomes should be expressed in terms of numbers and not merely as a set of verbal descriptions. These numbers could be classified in such terms as the age of the patient and stages of development. This method seems so sensible that it is difficult to see why it was resisted by the medical establishment.

Louis has been described as "the father of medical statistics"[24] and as "one of the pioneers of clinical statistics."[25] They take note, however, that some of his work can be faulted for drawing conclusions on insufficient evidence.[26]

Louis obtained an M.D. in 1813 and then practiced medicine in Russia, returning to Paris in 1820. He became associated with several hospitals in Paris. He stressed the importance of accurate data, including not only the symptoms of a patient's disease but, when the patient had died, detailed information—when possible—from an autopsy. It has been estimated that Louis performed 2,000 autopsies, devoting at least two hours to each one. Above all, he sought for quantitative information and developed what he called the numerical method. In 1827, he took a year off (in Brussels) where he used his freedom from hospital duties to analyze his data.

Among his publications, three stand out, two being his studies of tuberculosis and of typhoid fever, both of which depend on statistics and display his numerical method. His most celebrated publication was his refutation of the claims of a contemporary, F. J. V. Broussais (Paris, 1835), concerning the use of bloodletting in the treatment of pneumonia and other diseases.

Louis acquired an enthusiastic following, who accepted the doctrine of the numerical method. In 1832 his students formed a Société Médicale d'Observation. Louis was appointed "permanent President."

In 1835, Louis explained his new method in *Recherches sur les effets de la saignée dans quelques maladies inflammatoires, et sur l'action de l'émétique et des vésicatoires dans la pneumonie* as follows:

> Between the one who counts the facts, grouped according to their resemblance, in order to know what to believe regarding the value of therapeutic agents and him who does not count but always says "more or less frequent," there is the difference between truth and error, between something that is clear and truly scientific and something that is vague and without value—for what place is there in Science for that which is vague?[27]

In this classic work, Louis analyzed the effectiveness of blood-letting. At that time (1835) blood-letting, an ancient medical practice, was used extensively to treat patients suffering from pneumonia and other diseases. In the nineteenth century bloodletting was not performed surgically by making an incision in a vein, but rather was carried out by applying leeches or bloodsuckers to the patient's body. It has been estimated that a single treatment might require the application of as many as 50 leeches.

Louis's statistical evidence demonstrated the ineffectiveness of blood-letting. In asserting this, he placed himself in opposition to the French medical establishment. His primary opponent was a distinguished doctor, B. J. V. Broussais. In the late 1820s, Broussais himself used 100,000 leeches in a single year.[28] Today Broussais has an odd distinction: he is said to have been "the first physician to be destroyed by statistics."[29] Nevertheless, Louis was not able to change the outlook of the French medical profession all at once. Even those who agreed with his ideas in general understood that the numerical method yielded only probabilities, not certainties, and that his answers were at best statistical.

For Americans, Louis is an important figure because he had a number of students from the United States, particularly from New England. Sir William Osler estimated that Louis trained at least 37 American doctors.[30] Among the New England doctors who studied under Louis, the best known was Oliver Wendell Holmes, Sr., father of the jurist Oliver Wendell Holmes, Jr., and author of *The Autocrat of the Breakfast Table*.

Louis's influence on American medicine went beyond merely training some American students. His major publications were translated into English and published in American editions by Bostonian doctors: Henry I. Bowditch produced a revised translation of Louis's 1835 work on phthisis (tuberculosis) and in 1841 translated his study of typhoid. G. C. Putnam translated Louis's treatise on pneumonia in 1836, and in 1839 George C. Shattuck translated his tract on yellow fever.[31]

James Jackson, Sr., who was introduced to Louis's method by his son, hailed the Frenchman as the man he had been searching for, for 35 years: a medical investigator who actually practiced the Baconian ideals of measuring, weighing, and numbering. Louis alone had taken the gigantic step necessary actually "to pursue the method of Bacon thoroughly and truly in the study of medicine."[32]

Louis's research was publicized in American medical schools by Elisha Bartlett's popular *Essay on the Philosophy of Medical Science* (1844, 1852, 1856). Bartlett stressed the importance of Louis's numerical method, saying, "It is only by the aid of these principles [of observation, statistics, and mathematics], legitimately applied, that most of the laws of our [medical] science are susceptible of being rigorously determined."[33] The book was favorably reviewed by two of Louis's students, Josiah Clark Nott of Mobile and Alfred Stille of Philadelphia.[34] Stille wrote to George C. Shattuck of Boston that Bartlett's views were "those of my own medical creed, & yours, & of all of us who have been brought up in the school of . . . Louis."[35] Osler hailed Bartlett's *Essay* as a "classic in medical literature."[36]

Coleman, on the other hand, finds that Louis's actual "grounding in numerical data" was "slight." For example, in his tests of the efficacy of blood-letting, he "used only two series of cases, one of 78 patients (of whom 28 died) and another of 29 patients (4 of whom died)."[37]

Louis's response to such criticism was that he required both clear and simple facts, and reasoned generalization based uniquely on those facts. As he wrote in 1837 to one critic, Jean Cruveilhier, "all [knowledge] comes from experience, it is true, but experience is nothing if it does not form collections of similar facts. Now, to make collections is to count."[38]

NEW USES FOR NUMBERS: INNOVATIONS BY CONDORCET AND LAPLACE

The late eighteenth and early nineteenth centuries witnessed tremendous advances in the mathematical theory of probability, the mathematics used to analyze statistical data. This new science ultimately gave rise to a variety of spin-off disciplines that altered our daily life. Two contributors to the development of the theory of probability were Condorcet and Laplace.

Antoine de Caritat, Marquis de Condorcet (1743–1794), was a mathematician and a social theorist. He was a friend of Benjamin Franklin's and of Thomas Jefferson's, both of whom had contact with him in his role as secretary of the French Academy of Science.

Like Lavoisier, Condorcet was an enthusiastic supporter of the French Revolution in its early stages. But as the Revolution turned into the Terror, Condorcet, like Lavoisier, was condemned and died on the guillotine. Yet Condorcet, while awaiting certain arrest and inevitable death, wrote one of the most optimistic books ever written about the future of mankind. This philosophical masterpiece, *Esquisse d'un tableau des progrès de l'esprit humain*, was published posthumously by his wife.

Pierre-Simon, Marquis de Laplace (1749–1827), was the most important mathematician since Isaac Newton. He was a gifted contributor to both pure and applied mathematics, and it was he who transformed Newton's "rational mechanics" into "celestial mechanics" with his treatise *Mécanique céleste* (1799–1825). Laplace made many important contributions to probability theory and wrote a very influential book on the philosophy of probability, *Théorie analytique des probabilities* (1812).

Laplace applied the theory of probability to judicial voting. Whereas in the British system of law, a unanimous vote of a jury of 12 was required to establish guilt, in France a simple majority would suffice. Laplace believed that the British system implied a bias toward the protection of society and that the French system was unjust to the accused. He recommended a compromise, requiring the agreement of 9 out of 12 jurors, which he thought would provide a better system of justice.

The history of the development of the theory of probability is beyond the scope of this book. Our interest here is in numbers, and in this field Condorcet has been noted for his use of the new science of probability to analyze the voting process. But perhaps his most important contribution was to have made Laplace aware of this problem.[39]

So in place of a discussion of probability, we must turn our attention next to the politically even more consequential emergence of the new science of statistics.

6 A DELUGE OF STATISTICS

TABLES GALORE

During the first decades of the nineteenth century, the concern with numbers reached such proportions, especially in France, that a historian might describe this situation as a deluge. Tables of numerical data relating to problems of statecraft and the human condition abounded. Many such accumulations related to medicine and public health,[1] while others contained data on marriages, births, deaths, and the occurrence of crimes. *Recherches statistiques sur la ville de Paris et le departement de la Seine* was a notable series of publications concerned with statistics for the city of Paris and the "department" of the Seine. (A French "department" is similar to an American state or county.) Designed as a series of annual reports, the inaugural volume (for 1821), though late in being published, was so heavily used that a second edition was produced in 1833. A main concern of the *Recherches statistiques* was demographic: the population of the different districts and the changes that had occurred in those distributions. There were also tables of such data as marriages, deaths, and births. A lengthy introduction was geared to mathematically literate readers.

This publication was produced under the direction of Joseph Fourier, one of the ablest mathematicians of that time. He is particularly remembered for his contributions to the mathematical theory of probability, the underpinning of statistics. Readers who have done even a little work in applied mathematics, quantitative science, or engineering know Fourier for the "Fourier series," a mathematical tool which he developed.

Another serial publication went beyond the city of Paris and the department of the Seine. Called *Compte général de l'administration de la justice criminelle* (Account of the Administration of Criminal Justice), this statistical publication had a wide scope, extending over the whole of France, and made available to researchers the numerical data being accumulated by the Department of Justice.

These volumes consisted almost entirely of tables of data, listing in meticulous detail the occurrence of crimes in two major categories: crimes against persons and crimes against property. The crimes against persons include, in this order: cutting and maiming; murder; assassination (defined as "murder premeditated"); rebellion; rape and assault with intent to rape; rape on children; infanticide; false witnessing and bribery; cutting and maiming parents or guardians; poisoning; criminal conspiracy; crimes against children; parricide; abortion; bigamy; contempt of court and its officers; begging ("accompanied with violence"); political offences; threatening; breaking prison; breach of the sanitary laws; castration; false witnessing in civil cases; violation of public decency; forfeiture; and slave-trade.

A similar tabulation of crimes against property includes, again in order: robbery ("differing from the following"); robbery in dwelling-houses; fraudulent offences ("differing from the following"); forging in commercial documents; robbery on the highway; fraudulent bankruptcy; burning of buildings; sacrilege; frauds under false pretenses; counterfeit coin; exaction and corruption; extortion of signatures; destruction of movable or immovable property; pillage and destruction of grain; burning of various objects; counterfeiting seals; pillage and destruction of furniture; suppression of titles and deeds; forgery of bank-notes; defrauding the public treasury; smuggling; breaking open of sealed things; loss of a ship by negligence of the pilot; barratry; and abuse of a blank signature.[2] These lists are striking for their level of detail and for the fact that the order of categories seems completely arbitrary. There is, however, a simple reason for the arrangement: the crimes are listed in descending order of frequency.

Other tables list the commission of crimes by age of the offender, by sex, by season of the year, and by "department." What possible use could such tables serve? In the introductory essay to the first volume of the series (1827, containing data for the year 1825), it is wisely pointed out that one cannot "draw any certain conclusions from the data for a single year." The introduction promises, however, that the eventual correlation of the data from several years will serve to make plain the circumstances which "augment or diminish the number of crimes."

For us, these data have a special importance in a story of social numbers because they furnished the data for the numerical analysis of society by two brilliant researchers, A. M. Guerry and Adolphe Quetelet.

GUERRY'S STUDIES OF CRIME

André Michel Guerry, a wealthy lawyer interested in statistics, was one of the first researchers to make use of the statistics on crime published by the Judiciary Department of France. As he studied the tables of crimes, he was struck by the regularity with which, year after year, various kinds of crimes were committed in the country. One subject that particularly fascinated him was the numerical incidence of crime in relation to levels of education. He gauged the level of education by reference to literacy, using data on the number of literates and illiterates among recruits for the French army. These numbers showed that, contrary to all guesses and statements that had been made, regions with the highest levels of education displayed the highest incidence of crime.

Guerry's findings, published in 1833 in a large format volume entitled *Essai sur la statistique morale de la France*, won a prize from the French Academy of Sciences. The award made special note of the fact that Guerry used elaborate diagrams and charts, including a map of France divided into regions of different degrees of shading, to indicate his numerical findings. Guerry's work was made available to English-speaking readers in Henry Lytton Bulwer's study of France which reproduced the maps and charts. This book had wide circulation, being reprinted a number of times in both England and America.

One of the results of Guerry's work was the introduction of the expression "statistique morale" which is badly translated as "moral statistics." Guerry's statistics dealt with such social pathologies as alcoholism, prostitution, murder, suicide, and damage to property. Thus a much more appropriate name, as Ian Hacking once suggested, would be "immoral statistics." "Moral statistics" is no longer current in English usage, but the term is still used widely in France.

According to Bulwer, Guerry's work was "more especially remarkable on this account—that it bowls down at once all the ninepins with which late statists had been amusing themselves, and sets up again many of the old notions, which from their very antiquity were out of vogue." For example, according to Bulwer:

> Some very wise persons have declared that crimes depended *wholly* upon laws; others that they depended *wholly* upon,

what they called, instruction [that is, education]; while a few, with a still falser philosophy, have passed, in their contempt for all existing rules, from the niceties of calculation to the vagueness of accident, and insinuated, not daring to assert, that vice and virtue are the mere "rouge et noir" of life, the pure effects of chance and hazard.

It was against "these champions" that Guerry entered the field. "Of all the marvellous calculations ever yet published," Bulwer wrote, "this calculation is perhaps the most marvellous." The reason is that "whatever the basis on which the computation is made, it is not a whit the less wonderful that it should in six successive years give an almost similar result." And he noted that this was "not in one species of crime—not in one division of France, but in all the divisions of France, and in each distinct class of crime!"

One subject of particular interest to Guerry was suicide. (At that time suicide was still addressed as a crime.)[3] He collected evidence relating to suicide and displayed his findings in a way that dramatized his research. He made a collection of what Bulwer called "the different sentiments uppermost in the minds of different individuals at the time when they have deprived themselves of existence." The information in the table came from papers found on the persons of the deceased. He also used graphs to show an otherwise unsuspected correlation between the method of self-destruction—the noose or the pistol—and the age of the victim. According to Bulwer, a person who didn't have numerical data "would hardly imagine that the method by which a person destroys himself is almost as accurately and invariably defined by his age as the seasons are by the sun." Thus, "the young hang themselves; arrived at a maturer age they usually blow out their brains; as they get old they recur again to the juvenile practice of suspension."[4]

Guerry produced a second statistical study, *Statistique morale de l'Angleterre comparée avec la statistique morale de la France* (1864). This work, like its predecessor, was awarded a prize by the French Academy of Sciences. Once again Guerry was especially honored for his "atlas" (as he termed it) of elaborate charts and diagrams. These diagrams were presented in various colors, either printed in color or tinted after printing. Copies of this work, with both text and plates, are very rare. This work did not have the pioneering impact of his earlier study of France.

Several publications record that Guerry invented a machine which he called an "ordonnateur" or an "ordonnateur statistique."[5] This was apparently a device to aid Guerry in his extensive tabulation and analysis of numerical data. Then, in 1990, Ian Hacking noted that the name of this machine, "ordonnateur," was very much like the French name ("ordinateur") introduced by IBM-France for the computer.

In the mid-1950s, when the name "ordinateur" was introduced, the computer was known in France as a "computeur" or even "calculateur." The French found both terms offensive. First of all, "computeur" is a hybrid, a Frenchified English term and as such an example of what French scholars in derision call "Franglais." Second, the function of a computer had already surpassed mere computation or calculation.

IBM was particularly concerned lest they offend French sensibilities, so they were eager to find or invent a French name for the computer. To this end, in 1955, they commissioned a professor of language at the Sorbonne, Jacques Perrot, to produce a suitable French name for their product.[6] Professor Perrot came up with "ordinateur," a name that continues to be used to this day.

It is no longer possible to find out whether Professor Perrot was acquainted with the writings of Guerry and may have adapted Guerry's "ordonnateur" as the source of his own "ordinateur." In any event, the fact is unchangeable that Guerry's machine had almost the same name as the term used today throughout France for the computer.

But fate was not kind to Guerry. Although Guerry made extensive studies of suicide, Emile Durkheim does not mention his name in his classic work on suicide. Eclipsed by the brilliance of Quetelet, to which the next chapter is devoted, Guerry is barely mentioned in histories of statistics.

7 STATISTICS REACHES MATURITY: THE AGE OF QUETELET

NUMBERS, NUMBER SCIENCE, AND JOYCE'S *ULYSSES*

The publishing history of one of the most celebrated artistic creations of the twentieth century, James Joyce's *Ulysses*, shows in a dramatic way how a science associated with numbers invades life's activities in wholly unsuspected ways. Although *Ulysses* is widely distributed today, this book was long considered to be so "dirty" that it was completely banned in both Britain and the United States. That meant that the book could not be printed, sold, or imported. During the 1920s and early 1930s, travelers returning from Europe or Latin America routinely had their personal luggage searched, not only for alcoholic beverages but also for copies of *Ulysses* (along with D. H. Lawrence's *Lady Chatterley's Lover*).

This situation changed dramatically in 1933, when the ban on *Ulysses* was lifted by a decision of Judge John M. Woolsey, a U.S. District Judge, who ruled unequivocally that "*Ulysses* may ... be admitted into the United States." Judge Woolsey's decision has achieved a certain amount of fame because it was printed as a foreword to the Random House edition of *Ulysses*, the first legitimate American edition, and then reprinted again and again in the "Modern Library" series. I remember reading the text of this decision only many years after it had been made, at a time when I had completed my graduate studies and was already deep in my career of historian of mathematics and science. The reason that I had never seen this decision was that I was one of those fortunate few who had been able to obtain a "bootleg" copy before the ban had been lifted. It was one of those copies printed in Paris, bound in blue cardboard. And so it was only years later that I encountered the Random House edition and read Judge Woolsey's decision with his reasons why *Ulysses* would not undermine the morals of our country.

I learned later on, from some colleagues who were students of Constitutional law, that Judge Woolsey's decision was considered a landmark in legal history. His ruling that *Ulysses* could be published and sold in the United States did more than merely allow James Joyce's great work to be legally distributed in America. He took a great leap forward in the constant struggle against the dark forces of censorship. Many writers, publishers, and readers have hailed this decision, rejoicing that at last James Joyce's masterpiece could be permitted to circulate. This work, generally considered to be one of the literary masterpieces of modern times, was at last to be made accessible to readers. There has, however, been some serious concern that Judge Woolsey opened a veritable floodgate of pornography in which we have been drowning ever since.

Judge Woolsey's decision was notable in many ways. Not least was the fact that in writing it, he at times adopted a kind of poetic style aptly suited to the lyrical cadences of Joyce's *Ulysses*. For example, the Judge took note of factors that lessened the purely pornographic aspect of certain episodes. Noting the "recurrent theme of sex" in "the minds" of Joyce's characters, the Judge said readers should remember that "[Joyce's] locale was Celtic and his season Spring." Of course, the Judge remarked, *Ulysses* does contain some "old Anglo-Saxon words," which have been "criticized as dirty," but these words—he said—are "known to almost all men and, I dare say, to many women." Basically, Judge Woolsey based his final judgment on a legal issue of obscenity, whether this book tends "to stir the sex impulses or to lead to sexually impure and lustful thoughts."

As I studied Judge Woolsey's decision, I was not particularly interested in the legal issues involved. Rather, looking at his text from the point of view of numbers and the various sciences based on numbers, I was struck by the fact that Judge Woolsey did not ask whether this book might sexually stimulate *any* reader or even *most* or *some* readers; nor was he primarily concerned with what effect a reading of the book might have on people of different ages or temperaments. Rather, he based his decision on what he assumed might be the effect of the book "on a person with average sex instincts—what the French would call *l'homme moyen sensuel.*" That is, the Judge considered the effect on a person of average sensibility. Such a person, he explained, plays in this kind of enquiry the same "role of hypothetical reagent" as does "the 'reasonable man' in the law of torts" and "'the man learned in the art' on questions of invention in patent law."

Judge Woolsey was not concerned with the historical significance of this concept of "l'homme moyen." He showed no awareness that this was a mathematical concept that had come into prominence in the nineteenth century with the numerical analysis of social phenomena.

Legal scholars have been concerned with the validity of this concept of a "hypothetical reagent," but they have not known the historical and numerical significance of the type of person on whom the Judge based his decision. Nor have the lawyers who have discussed Judge Woolsey's decision been aware of the nature and historical context of the concept invoked by him—a "person of average sensibility." It is not generally recognized that the concept of such a person (an "homme moyen sensuel") was actually invented only several decades before the birth of James Joyce and was a cornerstone of a wholly new science of applied numbers, the science of statistics applied to problems of society. Judge Woolsey's fictitious being, an "average" man or woman, was the intellectual creation of Adolphe Quetelet, one of the most innovative and influential thinkers of the nineteenth century.

QUETELET'S WORLD OF NUMBERS

In the world of numbers, the middle decades of the nineteenth century were a crucial period of development. Governments became increasingly interested in counting their citizens and collecting numerical data about them. National statistical societies sprang up, and in 1853 the first of many international statistical congresses met.

The driving force behind this international movement was Adolphe Quetelet, who has been called the "powerhouse of the statistical movement," the "greatest regularity salesman of the nineteenth century."[1] Westergaard found that during the "Era of Enthusiasm" (1830–1849), "Quetelet overshadowed most statistical contemporary authors by his brilliant style, his vivid imagination and his abundance of ideas."[2] Additionally, "in the period concerned, he is the central figure in statistical literature," and "his contributions show both the strength of this period as well as its weaknesses."[3]

Although Quetelet's name will be unfamiliar to most readers, he certainly merits inclusion among "the movers and shakers" of the nineteenth century. His influence was not in a class with that of Charles Darwin, Karl Marx, or Sigmund Freud. Yet he did influence our thought to a remarkable degree. We still pay him honor when we use such concepts as his "average man," even if the sense of such words has shifted with time. He merits a high place in the pantheon of statistical thought.*

Adolphe Quetelet (pronounced Kettle-lay) was born in 1796 in the Belgian city of Ghent, at a time when Belgium was part of Holland. Although Quetelet demonstrated real ability in mathematics, his intellectual interests primarily focused on the arts. He wrote poetry and drama, was co-author of an opera, and also developed some skill as a painter. This artist's concern for the human figure was later to prove a valuable asset when he became concerned with the statistics of growth, height, weight, chest size, and so forth. Eventually a friend persuaded him to shift from the arts to mathematics, but he continued to write poetry until he was in his thirties.

Quetelet became a student at the newly created University of Ghent. In 1819 he obtained a doctorate in mathematics, the first such doctorate awarded by the new university. Then, at the age of 23, he was called to Brussels as a teacher of elementary mathematics at the Athenaeum. His students were apparently on the "adult education" level. In 1820 he was elected a member of the Belgian Académie Royale des Sciences et Belles-Lettres.

Quetelet taught a variety of courses on mathematics and the sciences. Two among them may particularly attract our notice. One course was on probability. Quetelet later based a popular book on these lectures.[4] The other course was on astronomy. Both subjects were important in Quetelet's eventual professional career.

In the 1830s, Quetelet for a short time tutored the princes

*In English, Hankins ([1908], 1968) is a valuable source for the life of Quetelet and also contains many extracts translated into English. An extremely valuable source is the article on Quetelet in the *International Encyclopedia of the Social Sciences*. Many of my ideas concerning Quetelet and his importance have been influenced by the writings of Paul Lazarsfeld and by conversations with him concerning Quetelet.

Ernest and Albert of Saxe-Coburg and Gotha in mathematics. Even after they had left Belgium to attend school in Germany, he continued their lessons by correspondence. His letters to them were later published as *Letters Addressed to H.R.H. the Grand Duke of Saxe Coburg and Gotha.* Ernest, the older brother, became Grand Duke; the younger brother, Albert, married Queen Victoria in 1840. Albert, like his brother, was greatly interested in science and technology and was likewise much influenced by Quetelet's teaching. In 1860, Albert delivered the keynote address to the fourth meeting of the International Statistical Congress in London.[5]

Quetelet became active in a movement to establish a Belgian national observatory. When the government finally decided to proceed with this project, Quetelet became the first director. At that time, the Royal Observatory at Paris was one of the world's leading centers for astronomical work, and the authorities decided to send Quetelet to Paris for several months to learn about the latest tools and techniques of astronomy and the organization of an observatory.

These months marked a turning point in Quetelet's career. In Paris he came into contact with Laplace, Poisson, and Fourier, three mathematicians who had been developing the mathematical structure of statistics. Under their tutelage and inspiration, Quetelet developed a consuming passion for statistics. He became convinced that statistics provided an insight into human behavior and the understanding of society, and thereafter he devoted his life to the production of statistical information and the interpretation of the numbers. He is held to be the founder of quantitative social science, of statistically based sociology.[6] It is difficult, as Theodore Porter has observed, to find out exactly how much contact Quetelet actually had with Laplace, Fourier, and Poisson.[7] But without doubt he was inspired by their teachings. It was their work that he had in mind when he insisted that statistics required more than a mere collection of numbers and had to be interpreted by the new mathematics of probability.

Laplace's influence is evident in Quetelet's reference to human intentions in terms of "perturbations," transferring to social analysis ("méchanique sociale") a basic concept of Laplace's celestial mechanics. In celestial mechanics, the motion of a celestial body—a planet or a planetary satellite—is caused by gravitational forces acting according to Isaac Newton's law of universal gravity. A simple example is the system of our Moon revolving around Earth according to principles of physics and the law of gravity. But the Moon's motion is not simply caused by the attraction of Earth; it is also affected by the gravitational pull of the Sun. This complexity produces deviations from a simple orbit, which are technically known as perturbations.

In 1830 Belgium attained its independence from Holland, and thereafter Quetelet became the director of the Commission Centrale de Statistique, a position he held until his death in 1874. A contemporary observer said that the "success of the Commission, thanks to Quetelet, was so great that many nations . . . hastened to found a central commission of statistics patterned after his."[8]

Quetelet had a direct influence in countries other than his own. For example, while serving as delegate to the British Association for the Advancement of Science in 1833, he was active in establishing a section on statistics, suggesting to Charles Babbage the formation of a national statistical society. Babbage agreed and there came into being the London Statistical Society, now the Royal Statistical Society. When the American Statistical Society was founded in 1839, the first item of business at the second meeting was to elect Quetelet an honorary member.

Quetelet's importance on the world stage is apparent in his abundant communication with major political and scientific figures of his day. Some 2,500 correspondents have been identified, including Goethe, Gauss, Faraday, Ampére, von Humboldt, Wheatstone, Joseph Henry, and James A. Garfield, then a member of Congress who, when he became president of the United States, wrote to Quetelet about ways to improve the American decennial census.[9]

These examples leave no doubt concerning Quetelet's importance and, indeed, there are many books and articles about him and his career as a statistician. Here we are not so much concerned with Quetelet's contributions to the science of statistics as with his role in pointing out the importance of numbers and number science in the formation of the new sciences of society.

THE BUDGET OF CRIMES

One of Quetelet's astonishing discoveries was the regularity of crime. His discovery was based on a critical study of successive annual reports of the French Ministry of Justice (*Compte général de l'administration de la justice criminelle en France*). With astonishment and shock, Quetelet realized from the reports that the numbers of crimes repeated themselves season after season and year after year: We "pass from one year to another," he wrote, "with the sad perspective of seeing the same crimes reproduced in the same order and calling down the same punishments in the same proportions." He could not help seeing this situation as a "sad condition of humanity!"[10]

"The part of prisons, of irons, and of the scaffold," Quetelet observed, is as "fixed" as "the revenue of the state." The dreadful fact seemed to be that we "can enumerate in advance how many individuals will stain their hands in the blood of their fellows, how many will be forgers, how many will be poisoners, almost as we can enumerate in advance the [number of] births and deaths that should occur."[11]

This *cri de coeur* appears almost word for word in other publications of Quetelet's. He clearly found great significance in his discovery. For example, on the title page of his study of the tendency to crime he stated that "there is a budget which is paid with frightening regularity, it is that of prisons, hulks [prison ships], gallows." This identical sentence occurs also in the text of the book and again in a footnote. He added that this was a situation we should try to alter.

Quetelet's statement interests us in a number of ways. First, and obviously, the existence of these regularities is important to any consideration of determinism and free will. If committing a crime is a free act of the will, how can there be the observed regularities in the numbers? Furthermore, these regularities suggest that in some way the nature of human society induces with great regularity the occurrence of crimes.

Consider the following example from his 1831 book *Research on the Propensity for Crime at Different Ages.*[12] From the publications of the *Compte général de l'administration de la justice criminelle en France*, Quetelet collected data for the four years before 1830, categorized the data according to crimes against persons and against property, and drew from the data remarkable conclusions. The data were consistent from year to year, whether they reflected the total of the accused or the ratio of crimes against property to crimes against persons. In Quetelet's words:

> During the 4 years which preceded 1830, one counted in France 28,686 defendants before the assize courts, that is to say, about 7,171 individuals annually, which gives one accused for 4,463 inhabitants, taking the population at 32 million souls. Moreover, out of 100 defendants, 61 were condemned to penalties more or less serious. . . . It becomes very probable that for a Frenchman the odds are one against 4,463 that, in general, he will be indicted during the course of one year;

moreover, the odds are pretty nearly exactly 61 against 39 that he will be condemned as soon as he finds himself indicted. These conclusions are supported by the numbers which [table 7.1] plainly shows.[13]

Quetelet used this evidence to demonstrate the following point:

We know very well that every year 7,000 to 7,300 individuals are brought before the criminal tribunals, and that 61 out of 100 are regularly condemned. . . . This kind of budget for the scaffold, the hulks [prison ships], and the prisons is paid by the French nation with a regularity doubtless much greater than is the financial budget."[14]

There was another disturbing aspect to Quetelet's discovery. He could predict with accuracy the number of crimes against persons and against property, but he could not predict who would be the perpetrators or the victims of these crimes. His data indicated that 1 out of every 4,463 French persons during this period of time would probably be accused of a crime, but no one could know in advance who that person would be. Evidently the new science of society would be unlike the science of Galileo and Newton. It would be based on statistical laws rather than simple deterministic laws.

Quetelet fully understood the implications of his discovery. He saw that a major disturbing feature of these regularities was the question they raised about punishment. If, as Quetelet recognized, some aspect of our society produces these regularly occurring crimes, on what grounds can we justify punishing criminals? As Quetelet wrote to Villermé in 1832, there was considerable doubt in his mind about individual responsibility for crime. "Society," he wrote to Villermé in 1832, "prepares the crime, and the guilty person is only the instrument."[15]

This sentiment was repeated almost word for word in

TABLE 7.1 SUMMARY CRIME STATISTICS FOR THE FOUR YEARS PRECEDING 1830

Years	Accused	Condemned	Inhabitants for one accused	Condemned from 100 accused	Accused of crimes against		Relationship between the numbers of both types of accused
					persons	property	
1826	6,988	4,348	4,557	62	1,907	5,081	2.7
1827	6,929	4,236	4,593	61	1,911	5,018	2.6
1828	7,396	4,551	4,307	61	1,844	5,552	3.0
1829	7,373	4,475	4,321	61	1,791	5,582	3.1
Totals	28,686	17,610	4,463	61	7,453	21,233	2.8

From A. Quetelet, Research on the Propensity for Crime at Different Ages, *translated with an introduction by Sawyer E. Sylvester.*

131

Quetelet's *Physique sociale* (1842): "Society prepares the crime and the guilty person is only the instrument by which it is executed." This same statement appears also in his *Treatise on Man*.

Another example from the data produced by the Administration of Justice showed another regularity for the same four-year period, a sameness in the instruments of manslaughter. (See table 7.2.) "Nothing, at first, would seem to have to be less regular than the progress of crime," Quetelet noted, because manslaughters are usually unpremeditated, committed following "brawls which arise without motive [and are] most fortuitous in appearance." The numerical data, however, indicate that this crime was committed at approximately the same rate and in the same manner year after year. Astonished, Quetelet remarked, "What can one say, then, about crimes that are reflected upon?"

Quetelet was not the only statistician to note the regular frequencies of crimes. These regularities were also noted by Guerry, a man much concerned about his reputation, which was overshadowed by that of the famous Quetelet. In fact there arose a bitter and unpleasant dispute over who had first made this observation. In 1831 Guerry wrote a long letter to Quetelet setting forth his own major findings on crime and on other subjects. Quetelet published the letter as a supplement to his *Tendencies to Crime* of 1833. In addition, Quetelet made explicit references to Guerry's data and work in other writings. The issue of who was first has been explored in meticulous detail by Frank H. Hankins and by Joseph Lottin, who agree in giving priority to Quetelet.[16]

Although both Guerry and Quetelet discovered the remarkable regularity of crime, they saw different significance in the discovery. In comparing Guerry's work with Quetelet's, Ian Hacking notes that Guerry was a lawyer whereas Quetelet was an astronomer. Guerry was concerned with reforms that needed to be made in the legislative system and with the possible correlation of crime and the level of education. One of Quetelet's goals was to improve the conditions of society that led to crime, but his major concern was to use the data on crime and his analysis of those data as a step toward the construction of a science of society, the science that he eventually came to call "social physics." According to Hacking, "Where Guerry was a man of meticulous fact, Quetelet was a man of vision, an astronomer who saw in the behavior of his myriads of fellow citizens regularities worthy of the stars."[17]

Authors generally write books according to a plan or outline, and Quetelet did compose several books in this way: for example, his popular book on probability, and his volume of letters on probability addressed to Prince Albert's brother, the Grand Duke of Saxe Coburg and Gotha, show the author following a plan. But others of his works grew by accretion. He routinely announced his findings in short reports to the Royal Academy. Then he would put together several of these notes to form an article which might be published in a journal in France or England. Finally, several of these articles would be collected to form a book.

His most important book was put together by this method. Published in 1835, it bore the title *Sur l'homme et le développement de ses facultés.* A subtitle proudly announced that the subject was "physique sociale" or "social physics." In a two-volume second edition (1869), "Physique Sociale" became the main title.

Quetelet could not generate all the data on human physique that he needed for his analyses, so he used whatever data he could find, sometimes from very unlikely sources. For example, somehow he found a study published in 1814 in the *Edinburgh Medical Journal,* a study reporting of the measurement of the chests of more than 5,000 Scottish soldiers. From this set of data he calculated the mean circumference and the variation around the mean, showing that the error law applied to human variability.[18] Many times he used collections of data that were sent to him.

TABLE 7.2 INSTRUMENTS USED IN MANSLAUGHTERS

	1826	1827	1828	1829
Manslaughters in general	241	234	227	231
Rifle	47	52	54	54
Pistol	9	12	6	7
Sabre, sword, and other permitted arms	8	2	6	6
Stiletto, poignard, and other prohibited arms	7	5	2	1
Knife	39	40	34	46
Stick, cane, etc.	23	28	31	24
Stones	20	20	21	21

Axe, pitchfork, and other cutting or piercing instruments	13	20	16	14
Hammer and other body bruising not otherwise specified	22	20	26	31
Strangulations	2	5	2	2
Thrown down or drowned	6	16	6	1
Kicks and punches	28	12	21	23
Fire	–	1	–	1
Unknown	17	1	2	–

From A. Quetelet, Research on the Propensity for Crime at Different Ages, *translated with an introduction by Sawyer E. Sylvester.*

135

Quetelet considered military records a source of consistent and reliable data. His use of such data is seen in table 7.3, based on data concerning recruitment or conscription for the French army. He also used mortality records as a source of data. Table 7.4 gives data from Quetelet's *Treatise on Man* based on such records.

Quetelet's writings on social physics are based on a fundamental belief that physical attributes (such as height, weight, strength), as well as intellectual attributes and, indeed, all aspects of human life and behavior, were quantifiable and thus subject to numerical analysis. He believed that the same regularities

TABLE 7.3 HEIGHTS OF 100,000 CONSCRIPTS TO THE FRENCH ARMY

Height (in meters)	Number of men
Under 1.570	28,620
1.570 to 1.598	11,580
1.598 to 1.624	13,990
1.624 to 1.651	14,410
1.651 to 1.678	11,410
1.678 to 1.705	8,790
1.705 to 1.732	5,530
1.732 to 1.759	3,190
Above 1.759	2,490
	100,000

From A. Quetelet, A Treatise on Man and the Development of his Faculties, *facsimile reproduction of the English translation of 1842 (Gainesville, FL: Scholars' Facsimiles & Reprints, 1969)*

TABLE 7.4 TABLE OF THE POPULATION OF BELGIUM

Age (in years)	Deduced from the table of mortality	Obtained directly by the census	Age (in years)	Deduced from the table of mortality	Obtained directly by the census
Birth	100,000	100,000	50	21,289	17,471
1	96,937	97,214	53	18,154	14,488
2	94,562	94,446	56	15,220	12,039
3	92,401	91,962	59	12,495	9,899
4	90,361	89,489	62	9,993	7,811
5	88,400	87,034	65	7,746	6,058
6	86,487	84,648	67	6,404	4,868
8	82,768	80,274	69	5,194	3,951
10	79,143	76,138	71	4,116	3,041
12	75,590	72,314	73	3,179	2,418
14	72,094	68,657	75	2,379	1,820
16	68,648	64,707	77	1,724	1,288
20	61,932	57,854	79	1,205	884
25	53,952	49,323	81	316	543
30	46,506	41,047	83	530	358
35	39,524	33,673	85	327	222
40	32,992	27,639	87	190	127
45	26,908	22,283	89	104	72

continued

90	76	50	96	8	6
91	55	33	97	4	4
92	39	25	98	2	2
93	27	18	99	1	1
94	19	13	100 & upward	1	1
95	12	9			

From A. Quetelet, A Treatise on Man and the Development of his Faculties, *facsimile reproduction of the English translation of 1842 (Gainesville, FL: Scholars' Facsimiles & Reprints, 1969)*

observed in the occurrence of crimes could be found in all aspects of our life as individuals and as members of social groups. This feature of human existence thus produced social facts that were the analogue of physical facts in the physical sciences. In fact, however, despite his accumulation of data from a large variety of sources, he was only to establish a regularity in three kinds of activity: crimes, suicides, and the rate of marriages.[19] In these Quetelet actually demonstrated that there are "regularities in man's 'moral' characteristics (those involving a choice of action)."[20]

Quetelet's *Treatise on Man* (later republished as *Social Physics*) contained tables of numerical data concerning every possible characteristic of life and social behavior. There are tables comparing the number of births to slaves and to free women at different ages. Another set of tables compares the number of children born to women who work in factories and those who have no employment outside the home.

The wealth of numerical social data by far exceeds normal expectations. Tables—often three or even four to a page—deal with almost every possible subject of human development and social behavior. Data came primarily from France, Belgium, and England, but some were from other European countries and even from America. The tables are based on criminal records, measurements of recruits or conscripts for military service, but also records of such social pathologies as drunkenness. Detailed numerical information is given on the marital status of the population—unmarried, widowed, married (for men and for women) at different ages.

Data are tabulated concerning the strength of men and women at different ages; the numbers were based on maximum weights that could be lifted with the right hand, with the left hand, and with both hands. A table is given of "the average weight of children of the lower classes," that is, the average weight of boys and of girls at different ages from 9 to 17, both those "working in factories" and those "not working in factories." There is even a tabulation of successful plays on the London and the Paris stages in terms of the age of the authors at the time of composition.

Each of these tables is the subject of a discussion interpreting the numbers and indicating their significance for a science of society. Just looking through the pages of Quetelet's *Treatise on Man*, or of the two volumes of *Sur l'homme et le développement de ses facultés* (Paris, 1835) of which it is a translation, we see an astonishing array of statistical numbers of social relevance. This was clearly not a philosophical or ethical treatise in the traditional manner. Here was a document in a new tradition of "statistique morale," an essay in (to use Quetelet's own favored designation) "physique sociale."

A central concept in Quetelet's social physics was the "average man," a person with average physical, intellectual, and other properties. Quetelet always stressed that he was referring to a particular population: for example, the inhabitants of a city, a state, or a nation. He was impressed by the fact that regarding some measurable feature, such as height or weight, most members of the designated population would present numbers clustering around the average.

Let us consider a hypothetical case, one not reported by Quetelet: the male population at a given beach on some appointed day. Let us suppose that the average height of this population is 68 inches. Most of these bathers would be either 68 inches tall or 67 or 69 inches tall. There would be very few, if any, giants (82 inches tall) or dwarfs (54 inches tall). A graph of the values would yield the familiar—and notorious—bell-shaped curve.

With the passage of time, Quetelet's concept of the average man underwent some notable changes. First, it was extended from physical properties (such as strength, height, and weight) to moral qualities (such as bravery). In this way, he believed, his social physics could attain the status of a "science" such as physics. Quetelet also introduced the notion of an "homme type," a typical man that might characterize a people, a nation. He came to believe that deviations from this type were accidents. Given such an homme type, he further believed, one could even reconstruct the whole population of the earth.

In the half-century following Quetelet, this particular concept of the average man came under attack, and in the form proposed by Quetelet, it all but disappeared from the literature of the social sciences. There remained, however, a significant use of the concept of averages in general, a fact that testifies to the importance of Quetelet and the value of his concept.

THE RELIABILITY OF STATISTICS

Quetelet knew that the worth of a statistical presentation depended on the accuracy of the data and the soundness of the method of analysis. In his volume of letters written to the Grand Duke, he specifically addressed this problem. "Most men," he wrote, "accept all statistical documents with equal confidence." Most people, he noted, do not check on the source of the data, the "mode employed in their collection," and "the number and value of the observations."

Each reader will know of cases in which statistical information is presented in a misleading way. A favorite example used by textbook authors concerns women students at Johns Hopkins University. This example is especially noteworthy because it shows how mere numbers may be misleading. At issue is a news item: CO-EDS MARRY FACULTY. This headline was explained as follows: "One-third of the women students at Johns Hopkins University during their first year married faculty members." This statement is literally true, but it creates an impression that is not supported by the actual circumstances. In fact, there were only three women students, and one of them married her professor.

The example clearly exhibits the misuse of statistics arising from "misleading statements." Quetelet pointed to another possible defect in statistics, more fundamental than misleading statements: the use of false numerical data. His primary example comes from American history. According to Quetelet, "During the War of Independence the United States carefully misrepresented the number of their population: they exaggerated considerably the number of inhabitants in maritime cities in order to put the enemy on the wrong scent." Clearly, Quetelet concluded, "no good appreciation of the American population could be founded on the documents of this period."

Indeed, Quetelet espoused caution with regard to the basic statistical numbers. This example of falsified data, however, has a curious aspect. Did this actually happen? Were the population numbers falsified as Quetelet declares? I put this question to several historians who specialize in American history, and none of them had ever heard of this supposed padding of the population data.

At bottom, Quetelet believed that statistics can serve two purposes—as documents to advance "the science" and as guides to the statesman. One of Quetelet's letters to the Grand Duke contained some practical guidelines to the use of statistics by an administrator. In Letter 45 he wrote that only by statistics could a statesman accurately decide what legislation was needed. Similarly, only by statistics could a statesman ascertain whether a particular piece of legislation had produced the intended result.

Quetelet's enthusiasm for numbers seems to have been unlimited. He sought numerical relations and rules in every possible aspect of human activity and in every natural event. Two examples will illustrate his zest. The first is the creativity of playwrights; the second is the blooming of lilacs.

Quetelet's studies of the creative power of playwrights were based on two objective sets of numbers: the age of the playwright and the quality of his plays. To avoid the pitfalls of subjective judgments, Quetelet assigned numerical values of a play's worth in terms of actual frequency of performance at two major theaters, one in London and the other in Paris.

Quetelet found that in "both England and in France, dramatic talent scarcely begins to be developed before the 21st year." Then, between 25 and 30, the talent "manifests itself very decidedly," and "continues vigorously until toward the 50th or 55th year." Then, "it gradually declines, especially if we consider the value of the works produced."

A second example of Quetelet's enthusiasm for applying numbers to many sorts of phenomena was his rule for predicting when lilacs bloom. He found that the lilac flowered when the sum of the squares of the mean daily temperatures, after the last frost, amounts to 4,274 degrees. The observer would have to note the date of the first appearance of leaves, then apply Quetelet's rule, and that date would mark the first appearance of flowers.

Indeed, Quetelet thought so well of his lilac rule that he

devoted a whole chapter (or letter) of his *Letters to H.R.H. the Grand Duke* to it, including extensive documentation to show how well the rule worked. He begged his readers to make extensive tests.

So far as I know, horticulturists have never taken this numerical excursus seriously. This would appear to be a case when Quetelet did not heed his own warning against an "immoderate use of numbers."

COMTE VERSUS QUETELET: SOCIAL PHYSICS, OR SOCIOLOGY?

One unanticipated consequence of Quetelet's choice of the name "social physics" was the invention of the name "sociology." The French philosopher Auguste Comte invented this name for the "scientific" study of society. "Sociology," is, in fact, an ugly word, pasting together a Greek root and a Latin root. It must not be thought, however, that Comte was simply an ignoramus, trained as a mathematician rather than a humanist. Unlike the probable inventor of other such hybrid names as *automobile* (from the Greek root "auto," meaning self, and the Latin "mobilis," meaning movable), he knew just what he was doing. He wrote, in fact, that he chose these two roots with a definite purpose in mind. The final part, "-ology," was used to establish a kinship with life sciences such as morphology, physiology, biology, and so on. The first part, "socio-," was chosen because it conveyed directly the idea of "society," derived from the Latin "socius," meaning ally or partner.

In the nineteenth century, when educated men and women knew Greek and Latin, it was obvious to them that "sociology" was a hybrid. We can see the difficulties of using this new word by turning to John Stuart Mill's *A System of Logic, Ratiocinative and Inductive* (London, 1843). Book 6 on "The Logic of the

Moral Sciences" discusses the methodology suitable for the social sciences. In the text itself, Mill uses both "sociology" and "the social science" as distinct from political science or political economy or history. In the beginning portion of Chapter 9, Mill originally wrote in his manuscript about "the Social Science . . . which I shall henceforth, with M. Comte, designate by the more compact term Sociology." On reflection, however, he could not so easily pass over this neologism, and so the published version discusses "the Social Science . . . which, by a convenient barbarism, has been termed Sociology."

Comte believed in a pyramid of knowledge, with mathematics as the base, physics built on mathematics, then chemistry, then biology, and finally the crown of knowledge, sociology. This scheme implied a historical hierarchy since physics depended on mathematics, just as a true or "positive" science of chemistry needed physics for its construction. In a sense Comte's scheme implied that to have a science of society, a sociology, a science of the behavior of human beings, it was first necessary to have a science of *individual* behavior, a "positive" science of biology.

Quetelet's approach was just the opposite. In his system there was no need to have a positive science of biology before creating a social physics. For him this new science was not reached by induction from the study of the behavior of individuals (sociology). Rather, Quetelet went directly by means of statistics to a science of *collective* human behavior.

Quetelet not only took a position poles apart from that adopted by Comte, but his ideas were also abhorrent to Comte because they were statistical and based on probability. Comte despised science based on probability. Rather than considering such studies as legitimate, he believed that the use of probability in science was merely a measure of ignorance, of a lack of true "positive" knowledge.

Apparently Comte had first conceived the new science of

society as a "social physics." Accordingly we can easily imagine Comte's state of mind on discovering that Quetelet used this name for a science of society very different from the one he had envisaged. And to make matters worse, "social physics" as conceived by Quetelet was based on statistics, on a foundation of probability—a method that implied an inexact or illegitimate science, that is, "chance."

Thus, the difference between Quetelet and Comte on this issue was not merely the choice of language, but actually the very nature of the form which the new science of society would take.

"Chance" or not, Quetelet's influence spread.

WHAT DID QUETELET ACCOMPLISH?

Among other achievements, Quetelet demonstrated that statistics and statistical analysis are not limited to treating problems of statecraft. He showed that a wide range of subjects related to society could be represented by numbers and could be analyzed by the new mathematical methods. Thus, although someone doing statistical analysis might still be called a "statist," the subject matter of statistics was much broader than problems of statecraft.

The influence of Quetelet's ideas spread throughout the sciences, even to the physical sciences. The two primary founders of the modern kinetic theory of gases, based on considerations of probability, were James Clerk Maxwell and Ludwig Boltzmann. Both acknowledged their debt to Quetelet. This example is especially interesting because historians generally consider the influence of the natural sciences on the social sciences, whereas in the case of Maxwell and Boltzmann, there is an influence of the social sciences on the natural sciences, as Theodore Porter has shown.[21]

The broad impact of Quetelet's ideas can be seen in an

extraordinary document, published as a book review in 1850 in *The Edinburgh Review*. Sir John Herschel, reviewing Quetelet's *Letters to H.R.H. the Grand Duke*, made clear the enormous change in scientific thinking resulting from Quetelet's work. Herschel (1792–1871) was eminently qualified to make this assessment, since he was an outstanding scientist (an astronomer) and was also an authority on scientific method and the philosophy of science.

Herschel wrote: "Men began to hear with surprise, not unmingled with some hope of ultimate benefit, that not only births, deaths, and marriages, but the decisions of tribunals, the results of popular elections, the influence of punishments in checking crime—the comparative value of medical remedies, and different modes of treatment of diseases—the probable limits of error in numerical results in every department of physical inquiry—the detection of causes physical, social, and moral,—nay, even the weight of evidence, and the validity of logical argument—might come to be surveyed with that lynx-eyed scrutiny of a dispassionate analysis, which, if not at once leading to the discovery of positive truth, would at least secure the detection and proscription of many mischievous and besetting fallacies. Hence a demand for elementary treatises and popular exposition of principles, which has been liberally answered."[22]

But the "lynx-eyed scrutiny of a dispassionate analysis," and the "discovery of positive truth" that came with it, were not universally welcomed. There were soon influential voices being raised against them as an assault on humanity.

8 CRITICS OF STATISTICS

Thus far we have examined various episodes in the history of social numbers that elicited an enthusiastic reaction, or at least acceptance of, this extension of quantitative thinking. But concurrently, many abhorred the introduction of numbers into discussions of human affairs.

CARLYLE AND CHARTISM

The Chartists of the 1840s in England called for political reform to benefit the working class—universal male suffrage, equal electoral districts, payment for members of Parliament. Most of the English establishment opposed the movement and its goals as radical and unnecessary, citing government statistics that showed that the workingman had nothing to complain of. But the writer Thomas Carlyle (1795–1881) believed the Chartists had right on their side, that official statistics misrepresented the real condition of the people, and that the government used these statistics to prevent reform.

"Tables," he wrote, "are like cobwebs . . . beautifully reticulated, orderly to look upon, but which will hold no conclusion. Tables are abstractions." Even if statistics showed general prosperity, the workingman's "discontent, his real misery may be great. The labourer's feelings, his notion of being justly dealt with or unjustly; his wholesome composure, frugality, prosperity in the one case, his acrid unrest, recklessness, gin-drinking, and gradual ruin in the other,—how shall figures of arithmetic represent all this?"[1] Carlyle was far from alone in his conviction that statistics robbed their subjects of their humanity. His critique was soon amplified by a voice even more powerful than his own; indeed, by perhaps the most famous writer of his era.

DICKENS AND STATISTICS

Charles Dickens (1812–1870), strongly influenced by Carlyle, agreed and incorporated Carlyle's ideas, which appear in many of Dickens's works. In the preface to *A Tale of Two Cities*, Dickens expresses his respect:

> Whenever any reference (however slight) is made here to the condition of the French people before or during the Revolution, it is truly made, on the faith of the most trustworthy witnesses. It has been one of my hopes to add something to the popular and picturesque means of understanding that terrible time, though no one can hope to add anything to the philosophy of Mr. Carlyle's wonderful book.

Dickens opposed statistics on two grounds. He found that political leaders too often used statistics to block social legislation that would help the London poor and the factory workers. Additionally, statistics tended to concentrate on averages rather than individuals. Dickens saw himself as the spokesman for the individual and he took issue with the dehumanizing aspect of statistics that reduced human attributes to a set of impersonal numbers.

Of course, Dickens also used numbers when they favored a

position he held. Thus his two periodicals—*Household Words* and *All the Year Round*—would use numerical data to spice their articles even though the negative attitude toward statistics was evident to every reader. Thomas Carlyle also would use statistics if considerations of numbers helped him to make a point.

In the present context both men serve to alert us to the fact that the new numerical science cut both ways. Antagonism, for instance, is symbolized by the statement attributed by Mark Twain to Lord Disraeli, "There are lies, damn lies, and statistics." It is a curious fact of history that no one has been able to find this apothegm in Disraeli's novels, speeches, or correspondence and scholars now assume that this saying was invented by Mark Twain. Never mind! As the Italians say, "Se non e vero, e ben trovato!"

THE MUDFOG ASSOCIATION FOR THE ADVANCEMENT OF EVERYTHING

Among Dickens's earliest writings is a series of articles written for *Bentley's Miscellany* in the early 1830s when he was in his late twenties. These articles poke fun at the recently founded British Association for the Advancement of Science (BAAS), which held its inaugural meeting in York in 1831. The very name invited satire, since it implied that the advance of science would be a British affair and would be made by the scientists banded together in the new organization. To a critical layman like Dickens, the speeches at the meetings had an aura of pomposity and self-congratulation that invited satire.

Let it be noted that at least one British scientist also recognized the pomposity of the BAAS. James Clerk Maxwell, one of the most important scientists of the nineteenth century, could not help making fun of the British Association, which he playfully converted into "The British Ass." Its members became

"British Asses." In a serio-comic poem he addressed the members as "Ye British Asses, who expect to hear/ Ever some new thing."[2]

Dickens's parody purports to be a "Full Report of the First Meeting of the Mudfog Association for the Advancement of Everything." The three scientists whose opinions and activities are reported are "Professors Snore, Doze, and Wheezy." Among the famous savants in attendance were "Mr. Slug, celebrated for his statistical research." Dickens's heavy-handed humor is manifested in the names he gives to the learned men attending the meetings. The vice-president of the statistics section, for example, is "Mr. Ledbrain." Among the useless statistics was a report on the number of skewers of dog-meat sold in London per annum. So even from this early start Dickens makes his views concerning the "value" of statistical studies quite clear. He never wavered in this judgment.

"DEATH'S CIPHERING BOOK"

One of the bitterest condemnations of statistics published in Dickens's magazine *Household Words* was an article written by Henry Morley, a close associate. The article, published in 1855, was grimly titled "Death's Ciphering Book." There can be no doubt that as editor Dickens would never have published an article that did not express his own point of view.

"Death's Ciphering Book" purports to be an account of a meeting of manufacturers who were opposed to a recent government regulation that would require them to fence off dangerous machinery that might cause injury or death to the workers.

The argument against the government ruling was based on the fact that the number of such accidents was very small compared to the total number of workers in England. This was just the kind of argument that roused the ire of Dickens and Morley, who focused on individuals and not percentages, and on the fact that human beings actually suffered from contact with dangerous machinery. Morley seems to have enjoyed writing attacks on public figures who would quote statistics to preserve and exploit some condition of society. His argument was usually based on the assertion that no numerical calculation could take the place of "moral calculation." Thus, in "Death's Ciphering Book," Morley argued from the following principle: "As for ourselves, we admit freely that it never did occur to us that it was possible to justify, by arithmetic, a thing unjustifiable by any code of morals, civilized or savage."

Here is a sample of Morley's invective:

> We will not call it inhumanity—it is not that—but it is surely a strange illustration of the power of self-interest and habit, that a gentleman of high character, who well deserves all the respect attaching to his name, could think a point of this kind settled by the calculation, that four thousand accidents, great and small, yield only one to every hundred and seventy-five persons, and that the number of horrible deaths caused yearly being only forty-two—seven hundred thousand, divided by forty-two, gave a product of sixteen thousand and sixty-six, or, in round numbers, one in seventeen thousand.

Morley's point is made explicit by the numbers: a total of 4,000 accidents is equivalent to one accident for every 175 people working in factories and mills; 42 deaths is equivalent to 1 death in 17,000. Note that Morley's presentation of ratios and the arithmetic in the calculation is confusing!

Morley then asks:

> What if you were to carry out this method of arguing by products? There is a kind of death which the law seeks to prevent, although it is scarcely found to be preventable, and that

is by willful murder. Perhaps there may be about forty-two who suffer death in that way, annually, throughout Great Britain; and the population of the whole country is immensely greater than the population of the factory-world contained within it. Perhaps, also, there may occur in the year four thousand burglaries of greater or less moment, or some other number which would go certainly oftener than a hundred and seventy-five times into the whole population. Why, then, let it be asked, are honest men to be taxed for the maintenance of expensive systems of law and police when the per centage of burglary and murder, upon the sum total of men who are neither murderers or burglars, is represented only by such a ridiculous fraction as may be received at an aggregate meeting like the Manchester chairman's with laughter and applause? He spoke of a third of a man per cent. Burglary and murder together do not touch of a third of a man per cent, or anything approaching it. What right then has the home government to concern itself about such trifles as burglary and murder? This is the sort of argument to which we are reduced when the moral element is exchanged for the arithmetical.

Once again, Morley's expression of the calculation is confusing. By "product" he probably means "result," and by "dividend" he probably means "quotient."

CORRESPONDING DISDAIN

Dickens's disdain for arguments based on numbers and statistical averages is even displayed in his correspondence. For example, in a letter of 1864 to Charles Knight, one of the contributors to Dickens's *Household Words*, Dickens stated plainly and unambiguously that he was "against those who see figures and averages, and nothing else." Such people, according to Dickens, were "representatives of the wickedest and most enormous vice of this time." Through "long years to come," these statisticians will do "more to damage the real useful truths of political economy, than I could do (if I tried) in my whole life."

Such people had "addled heads." They "would take the average of cold in the Crimea during twelve months, as a reason for clothing a soldier in nankeen on a night when he would be frozen to death in fur." ("Nankeen" is the name of a light cotton cloth originally made in Nanking, China.)

Such statistically minded people, according to Dickens, would "comfort the laborer in travelling twelve miles a day to and from his work" by telling him that "the average distance of one inhabited place from another in the whole area of England is not more than four miles." Dickens added the comment, "Bah!"

Dickens's letter was a response to his having received a copy of Knight's book, *Knowledge Is Power.* According to Knight, Dickens "bore too hardly upon those who held that the great truths of political economy . . . were not insufficient foundation for the improvement of society." Knight feared that Dickens would set him down as a cold-hearted political economist. The issue of temperatures had been raised by a letter to *The Times* in which data concerning isothermal lines and meteorological tables had been used to show that the temperature at Sebastopol was only a little lower than at Paris and at London and a little higher than at Dijon. A later letter by *The Times* Crimea correspondent dismissed this nonsense as the work of a "philosophical idiot."

FACTS AND FIGURES: THE MESSAGE OF *HARD TIMES*

Published in 1854 and dedicated to Thomas Carlyle, Dickens's *Hard Times* is a savage, bitter novel. Dickens was 42 years old when this, his fourteenth novel, appeared, basically a broadside attack on an economic system that crushes the workers, destroys individuality, and glorifies the rule of head over heart. While it is an undisguised condemnation of the factory system—and succeeds as such—literary critics generally agree that it is one of Dickens's least successful works of art.

The theme is set in the opening pages, in which Mr. Thomas Gradgrind, a retired wholesale hardware merchant and the leading citizen of Cokeville, conducts a class in the local school. The quality of the school and the education it provides are made clear by the name of the schoolmaster, Mr. McChokumchild. In the opening sentences, Mr. Gradgrind sends his message, "Now what I want is facts." In this life, he repeats, "we want nothing but facts." And, of course, the simplest uncontestable facts are numbers.

Thomas Gradgrind is a "man of realities," a "man of facts and calculations, who proceeds on the principle that two and two are four, and nothing over." He always has in his pocket "a rule and a pair of scales," ready "to measure any parcel of human nature, and tell you exactly what it comes to." It is "a mere question of figures, a case of simple arithmetic."

Mr. Gradgrind even reduces the students to numbers rather than treating them as individuals. Thus he calls upon a girl as "girl number twenty." He doesn't refer to her by name or by her seat location (e.g., the girl in the back row). This is the introduction of Cissy, a major character who represents the values of Heart over Head. She can't even get herself to pronounce the word "statistics." The nearest she can get is "stutterings."

In the course of the novel, Gradgrind suffers two tremendous emotional blows.

The first concerns his son Tom. In a dramatic moment toward the end of the novel, this leading citizen of the town discovers that Tom is a thief. Tom has not only robbed the local bank but he has also cleverly arranged that an honest workman be blamed for the crime. "If a thunderbolt had fallen on me," said the father, "it would hardly have shocked me more than this."

Tom excuses his base conduct by referring to the principles of statistical determinism, as expounded by Quetelet. "I don't see why [you are shocked]," grumbled the son. "So many people are employed in situations of trust; so many people, out of so many, will be dishonest. I have heard you talk, a hundred times, of its being a law. How can *I* help laws? You have comforted others with such things, father. Comfort yourself!" We have seen in Chapter 7 that, as Quetelet wrote to Villermé in 1832, there was considerable doubt in his mind about individual responsibility for crimes. "Society," he wrote to Villermé in 1832, "prepares the crime, and the guilty person is only the instrument."

The second blow to Gradgrind concerns his beloved daughter Louisa. The stage is set for Louisa's tragedy when Bounderby, the local banker and mill owner, asks Gradgrind for his daughter's hand in marriage. Gradgrind's presentation of this proposal to Louisa takes place in his office, a room of his home lined with blue books, government reports giving numerical data on all sorts of political and social questions. Dickens describes it as a "stern room" and compares it to an observatory, designed to shut out the real world of Coketown and its inhabitants. On the wall there hangs what Dickens calls a "deadly statistical clock."

Gradgrind begins the conversation by remarking that he is confident that Louisa is not "romantic" or "impulsive," that her education has conditioned her to "view everything from the strong dispassionate ground of reason." To Gradgrind's surprise Louisa's reaction to the proposal is to ask, "Father, do you think I love Mr. Bounderby?" When Gradgrind gives an evasive answer, Louisa pursues the question. "Father," she now asks, "does Mr. Bounderby ask me to love him?" Gradgrind is at a loss how to reply. Love is not a quantifiable term in his equations. Finally he responds to Louisa's persistence with a parade of statistics on the number of successful marriages between people of different ages. These irrelevant data come from the "British provinces in India" and in Tartary and China. Three-quarters of such marriages, he reports, are successful and so "the statistics of marriage" favor Bounderby's proposal.

Gradgrind then extends the discussion to life expectancy and recent studies of the "duration of life" that have been made by the "life assurance and annuity offices." Gradgrind remarks that length of life "is governed by the laws which govern life in the aggregate." Louisa agrees to marry Bounderby. But she cannot end the discussion without taking note that the education and training she has received have not included the human emotions, that she has never had "a child's dream," never had a "child's heart." As Dickens will describe this education later on, it has been based on "the rule of Head over Heart."

Some time later, when the marriage has fallen apart, and when Louisa experiences a sensation which may be love, she returns distraught to her father's house. It has been storming outside and she appears soaking wet and in an obviously distressed state. Once again she has an intimate conversation with her father. It takes place in that same observatory, with the "deadly statistical clock" ticking at its inexorable pace. In her agony of spirit Louisa tells of her despair, she reminds her father that he has trained her "from the cradle" to be rational only, and then cries out, "I curse the hour in which I was born to such a destiny."

Louisa then asks him, "How could you give me life, and take from me the inappreciable things that raise it from the state of conscious death?" Specifically she bemoans the absence of what she calls "the graces of my soul," "the sentiments of my heart." In her distress she laments that her father never knew "that there lingered in my breast sensibilities, affections, weaknesses capable of being nourished into strength, defying all the calculations ever made by man, and no more known to his arithmetic than his

Creator is." Had her father not been limited by his statistical or numerical point of view, he would surely not have given her to the husband she now hates. This scene ends with Louisa, "the pride of his heart and the triumph of his system," collapsed in a heap on the floor.

Gradgrind's system of values has not prepared him for these disclosures of his favorite daughter. He tells her that the very "ground on which I stand has ceased to be solid under my feet." The "only support" on which he "leaned, and the strength which it seemed, and still does seem, impossible to question, has given way in an instant." He has been forced to admit that there is a "wisdom of the Heart" as well as a "wisdom of the Head."

Here is Dickens's rejection of a system of values that is based on numbers, on averages and statistics. This part of the novel concludes with Gradgrind's system in disarray, with a recognition that the rule of numbers is not enough for the good life, that the rule of the head must be supplemented by the rule of the heart.

9 FLORENCE NIGHTINGALE

Florence Nightingale merits a chapter in a book on social uses of numbers for two reasons. First, her use of statistics exemplifies a growing awareness in the nineteenth century of the practical importance of numbers in any discussion of public health and social reform. Second, she herself made contributions to the ways numerical information is presented.

Nightingale was a pioneer in using graphical forms to display statistical information. She was aware that many people are put off by numbers and cannot easily understand or accept the evidence of tables of numerical data. Today we are accustomed to graphical displays of data, but they have not always been in use. Nightingale did not invent this mode of presenting numerical information but she did make notable advances in it.

Florence Nightingale is generally known as the organizer and superintendent of nursing at the British military hospital at Scutari (Turkey) during the Crimean War. In this role, she was admired for her compassion for the sick and wounded soldiers committed to her care, and immortalized as "the Lady with the Lamp" in Henry Wadsworth Longfellow's poem, "Santa Filomela." She is widely known as the founder of the profession of nursing. Because of her accomplishments after her service in the Crimea, despite being a semi-invalid, she has been cited, along with Elizabeth Barrett Browning and Charles Darwin, as an example of "creative malady."[1]

Florence Nightingale was born in 1820 in Florence, Italy, the second daughter of wealthy English parents who were on a two-year tour of Europe. Her father, a man of leisure, had studied at Edinburgh and at Trinity College, Cambridge, being particularly interested in languages and philosophy. In the nineteenth century it was impossible for a woman to attend a university such as Oxford or Cambridge, so Florence's father, a Unitarian with advanced views on the education of women, taught his daughters himself. Under his tutelage Florence learned Greek and Latin; became fluent in French, German, and Italian; read widely in history and philosophy; and wrote essays as university students did.

From an early age, she was interested in numbers and in tabulating information. When the family traveled through Europe, she kept a notebook to record the distance traveled each day, the times of departure and arrival, and "notes on the laws, the land systems, the social conditions and benevolent institutions" of the regions they traveled through.[2] So, it is not surprising that at the age of 20 Florence expressed an interest in learning mathematics. Although opposed by her mother, who wanted her daughters to marry well and saw their education only as a means to this end, Florence studied mathematics with the help of a tutor.

Further, this multifaceted woman was a mystic, who several times in her life, she said, experienced the immediate presence of God and at the age of 16 experienced a first call to God's service: "God spoke to me and called me to His service," she later wrote in her diary.[3] The specific nature of the service was not made clear, just that she was to help her fellow human beings. On three later occasions she again had a direct communication with God:

before she went to become superintendent of the Harley Street nursing home; before she went to the Crimea; and after the death of her ally and friend Sidney Herbert.

Only gradually did it become clear to Nightingale that she was to become a hospital nurse. At that time, nursing was not a respected profession. Hospital nurses were generally "coarse" (that is, of doubtful morals) and ignorant women, lacking professional training, and often given to promiscuity and drunkenness. Florence's mother and sister vehemently, and naturally, opposed her plan, and her father, after consulting various doctors of his acquaintance about the practicalities, did not support it either. Florence, however, had no doubt that she was following God's command in trying to become a nurse.

At this time no institution existed in England where a young woman could receive professional training as a nurse. But the Catholics had a program. In fact, Florence considered becoming a Roman Catholic, and then a nun, so as to be trained as a nurse. Wiser heads persuaded her that this was no valid reason to convert. Eventually her parents allowed her to spend three months at a German institution (Kaiserswerth), a hospital and orphanage run by a Protestant order of Deaconesses. She supplemented this training with a brief apprenticeship in a French hospital run by a Catholic order, the Soeurs de la Charité, and she inspected as many schools and hospitals, in England and other countries, as she could. Along the way, she kept detailed notes of everything she saw and collected reports, regulations, and forms.

Nightingale's family continued to hope that she would give up her notions and marry. But she believed that the work she hoped to do would be impossible if she married, and so she refused two suitors. No, she had been called to the single life to be of service to God.

Finally, in 1853, Nightingale achieved a first step on her chosen path: she was appointed the (unpaid) superintendent of an institution for the Care of Sick Gentlewomen in Distressed Circumstances, in Harley Street, which Nightingale described as a nursing home for sick governesses,[4] in London. Sir Harry Verney, the grandson of Nightingale's brother-in-law, also Sir Harry Verney, relates the family tradition that she took this post "in the teeth of the fiercest opposition from her family." Her mother "stormed, lamented and had to be given sal volatile"; her sister "wept, raged, worked herself into a frenzy with hysterics, collapsed and had to be put to bed."[5] Her father, however, had made this position possible by giving her an allowance sufficient to live on.

A Committee of Ladies, who had some prejudices of the time, had hired Nightingale. As she wrote to a friend, the Committee had initially refused to allow her to "take in Catholic patients—whereupon I wished them good-morning, unless I might take in Jews and their Rabbis to attend them." Such conviction was rare, and the Committee quickly backed down, agreeing in writing to allow her to take in "all denominations whatever."[6]

During her one year in this position, Florence Nightingale greatly improved all aspects of the Harley Street operation. She trained the nurses, saved money on supplies, straightened out the accounts, reorganized the housekeeping and cleaning, and improved the food. She had hoped to establish a training school for nurses as well, but she was not able to accomplish this, to her bitter disappointment. But, one thing not considered so important at the time, she did keep systematic records of patient admissions, of categories of illness and outcomes of treatment, and of discharges and deaths.

In June 1854, British and French troops were sent to the Crimea, on the northern coast of the Black Sea, in support of Turkey in its dispute with Russia over Russia's demand to be given a protectorate over Orthodox subjects of the sultan. Even before there had been any actual fighting, bad sanitation led to disease among the troops: a goodly number of the expeditionary force came down with cholera, diarrhea, dysentery, and other disorders, and many died. In the initial battles in September and October of 1854 large numbers of English soldiers were wounded. After little or no emergency treatment they were removed by slow and ill-equipped ships to one of the military hospitals at Constantinople. They arrived weak, emaciated, and suffering from frostbite and dysentery as well as their wounds.

In truth, English military hospitals were completely unequipped to provide effective care, being filthy and vermin ridden, and lacking enough beds, tables, and chairs, and even bandages, tubs, towels, soap, and hospital clothing. There was basically no useful nursing care. Female nurses were not employed in British military hospitals at that time; nursing, such as it was, lay in the hands of old pensioners ("not of the slightest use," according to the *Times* correspondent), or by the sick and wounded themselves, helping each other.[7]

War correspondents reported in the English newspapers on the nightmarish conditions in the hospitals, and the public demanded that something be done. The Secretary at War, Sidney Herbert, knew Florence Nightingale, and wrote asking her to organize a group of female nurses who would be assigned to the army hospital at Scutari, across the Bosphorus from Constantinople. Herbert's letter to Nightingale crossed a letter from her to him, offering her services.

When Nightingale arrived in Scutari in early November, she found appalling conditions. The hospital, a former barracks, had been built over sewers with no outlet (essentially a cesspool). Foul air rose into the wards through open privies. Wounded soldiers lay on straw mats in overcrowded wards and corridors; the hospital was infested with fleas and rats; the only sheets were so coarse that the wounded preferred to be wrapped in dirty blankets. Laundry was done in cold water, which got nothing clean. Basic supplies were lacking. Food was scarce and bad.

From mid-November to mid-December of 1854, Nightingale later calculated, 15.5 percent of the treated cases died, many more from disease than wounds.

Nightingale's female nurses were the first, and being first, met opposition. Many medical officers opposed having women around, especially meddling outsiders and civilians. Undaunted, Nightingale gradually won acceptance for her nurses, and for herself.

The germ theory of disease had not yet been established. Even in later years, so pioneering a figure as Nightingale did not believe in infection by germs, but rather believed that diseases, or "conditions," were spontaneously generated in dirty or unventilated rooms or hospital wards. Surely, she opined, "morbid effluvia" from a sick person, unless removed by ventilation and cleaning, would reinfect the patient and prevent recovery.

Clearly abundant clean air, and cleanliness of both patient and nurse, were essential to good nursing care, and Nightingale based many of her reforms on that belief.[8] She immediately requisitioned 300 scrubbing brushes for cleaning the wards, and set up a laundry with boilers to provide hot water. Never imperious or setting herself apart, she also supervised the nurses she had brought and took night shifts herself.

Equipment and supplies had to be bought—she paid for much of this herself, or out of donations from the English public to a fund set up by the *Times*—and obtained such necessities as furniture, tableware, combs, bedpans, and clothing. Several extra kitchens soon provided better food.

Record keeping in the Crimea was pathetic. Previously even the number of patient deaths had not been accurately known; three different records were kept, giving different totals. Nightingale set this straight, keeping detailed daily records of admissions, wounds, diseases, and deaths. She reported continually to Sidney Herbert about the conditions she encountered, including incompetence, administrative confusion, overlapping authorities, and bureaucratic inertia, always urging reforms.

In January 1855 the mortality in all British hospitals in Turkey and the Crimea (excluding men killed in action) reached an annual rate of 1,174 per 10,000, calculated on the average patient population.[9] Of this number, 1,023 deaths per 10,000 were attributable to epidemic, endemic, and contagious disease.

In this same month, the British government fell, precisely on the issue of appointing a committee of the House of Commons to inquire into Crimean army conditions, and Sidney Herbert was forced to resign as Secretary at War, though he remained in Parliament, and Nightingale continued to report to him.

In February 1855 the mortality of cases treated in the hospital was 43 percent.[10] In late February the new government sent a "Sanitary Commission" to Turkey to look into the problems. With much broader power to take action than Nightingale had, they energetically set to work at sanitary engineering. They cleaned latrines and cesspits and flushed the sewers, removed the carcass of a dead horse from the water supply, ventilated the wards, and renovated parts of the buildings.[11]

A month later, the hospital death rate had dropped sharply, all owing to accurate calculations Nightingale had systematically collected. And the mortality continued to fall: later she wrote that during the first seven months of the campaign (September 1854 through March 1855), mortality among the troops was 60 percent per year from disease alone, a rate exceeding that in London during the Great Plague. But during the last six months of the war (April through September 1855), deaths among the troops dropped to only two-thirds of that among the troops at home.[12]

Neither the causes of disease, nor the way sanitation prevented it, were understood in the 1850s, but the evidence of the numbers was clear and indisputable. The health of British soldiers in hospitals depended on clean water and fresh air. Ergo, early Crimean mortality rates had been preventable, and thousands of British soldiers now lay "in their forgotten graves" who should not have died. Without reform, history would repeat itself.

But Nightingale also recognized that such syndromes reached all the way home to London. Indeed, all corners of the Empire cried out for reform.

SANITARY REFORM: THE EVIDENCE OF THE NUMBERS

When Nightingale returned to England in July of 1856, after the Crimean War had ended, she found herself world-famous, toasted all over England as a heroine. Here lay an opportunity, and Nightingale resolved to use her influence to improve sanitary conditions in the army. Immediately, she launched a campaign to get the government to appoint a royal commission to investigate the health of the British army. Once it had been appointed, she pushed for reform.

In this and later efforts Nightingale had the help of a group of sanitation enthusiasts, many of them comrades from duty in the Crimea.

These men—all were men—included Sidney Herbert (the former Secretary at War); John Sutherland (trained as a doctor, and one of the leading sanitarians of the day; he had led the Sanitary Commission to Scutari); Arthur Clough (a poet and activist, who was married to her cousin); and William Farr (1807–1883), a leading government statistician and a physician. These extraordinary men came to act on her behalf by what has been described as Nightingale's mixture of "flattery, nagging and cajolery"—a powerful combination.[13]

Nightingale needed the help of these men partly because a woman could not appear to play certain roles in such a campaign for reform, such as serving on an investigative commission, or even appearing before it as a witness. Also by now partly she was a semi-invalid. Having been dangerously ill in the Crimea, she remained in poor health for the rest of her life.

No one knows exactly the cause of her poor health. It has been suggested that she was suffering from brucellosis, sometimes called "Crimean fever."[14] Other suggest that Nightingale developed a psychological disability as a means of escape from the possibly overpowering influence of her mother and sister.[15] A recent suggestion by Hugh Small also cites psychological factors, a kind of shell-shock, flowing from knowledge that thousands of soldiers had fallen from simple ignorance.

Whatever the source of her illness, a time of rest and seclusion provided ideal conditions for her crowning creative work—mammoth studies of sanitation and health, produced when she was so ill that she could only tolerate one visitor at a time. These treatises must stand as one of the great monuments of the force of will under conditions of illness that in most people would simply cause all intellectual activity to cease.

At a dinner party in the autumn of 1856, Florence Nightingale met William Farr. Farr had been trained as a physician, but was not in practice. In the 1830s he had spent two years in Paris, where he attended Pierre Louis's lectures on the numerical method in medicine. During the next years he became concerned with problems of sanitation, and also became interested in statistics. Beginning in 1839 he worked as "Compiler of Abstracts" in the General Registry Office (where vital statistics were kept).

During his career in this office Farr produced, among many reports, the *English Life Tables* (1864), prepared for the use of actuaries making computations for life insurance and annuities. The life tables were based on registrations of births and deaths, and on census figures in England and Wales. The tables required immense numbers of computations, and Farr employed for the purpose a calculating machine based on the "difference engine" invented by Charles Babbage (1792–1871). The machine had been developed to a practical level, and a printing apparatus attached to it, by Edvard Scheutz and Bryan Donkin in 1856.[16]

Farr was convinced that statistics would provide the key to social reform and the improvement of health-care delivery services. His view of statistics was in harmony with the views of British statisticians of the 1840s, as expressed in the opening pages of the *Journal of the London Statistical Society*. "The Science of Statistics," the founders declared, differs from "Political Economy, because, although it has the same end in view, it does not discuss causes, nor reason upon probable effects." Statistics "seeks only to collect, arrange, and compare, that class of facts which alone can form the basis of correct conclusions with respect to social and political government."[17]

By the time that Nightingale met him in 1856, Farr was established as a major figure in statistics, author of a number of publications relating to public health. He was elected a fellow of the London Statistical Society in 1839 and was a founding member of the Social Science Association. He wrote with authority as a fellow of the Royal Society.[18]

Farr's part in Nightingale's campaign has been summarized by Diamond and Stone as follows: He was able to glean the greatest possible practical information from the limited official statistical sources; he could advise Nightingale on statistical procedures, and provide her with information from official sources, sometimes before such information was generally available; and he was in sympathy with her goal of sanitary reform.[19] "It will always give me the greatest pleasure to render you any assistance I can in promoting the Health of the Army," Farr wrote to Nightingale in 1857. "We shall ask your assistance in return in the attempts that are now being made to improve the Health of the Civil population."[20]

Working in the General Registry, William Farr had constructed from the vital statistics available to him life tables, showing death rates in civil life, by age group and other categories. When he met Nightingale he shared this information with her. She compared his figures with the death rate in army barracks and was appalled: even in peacetime the army mortality was nearly double that of civilians. She termed this situation "criminal."[21]

In response to pressure from Nightingale and her allies, including Farr, a royal commission of inquiry was established. The Secretary at War asked her to prepare a report of her experiences and observations in the Crimean campaign, and her views on sanitary reform generally.

Nightingale saw the importance of argument based on numbers. For this purpose, she compiled comparative statistical data on mortality rates, and on the rate of occurrence of different diseases, among British soldiers. In her report, "Notes on matters affecting the health, efficiency, and hospital administration of the British Army, founded chiefly on the experience of the late war," she showed that the bad health of the British army in peace was hardly less appalling than the mortality during the Crimean War.

In 1858 the commission issued its report on the health of the British army, "Report of the Commissioners appointed to inquire into the regulations affecting the sanitary condition of the army, the organization of military hospitals, and the treatment of the sick and wounded; with evidence and appendix." Nightingale's experiences in the Crimea, and the statistics she had gathered there, were both a major source for the report and a basis for the reforms the commission called for.

In response to the commission's report and to further pressure from Nightingale and her supporters, four subcommissions were established to carry out her reforms. The design of army barracks was improved, specifically the heating, ventilation, water supply, kitchens, and sewage disposal. An army medical school was established, and the procedures for gathering medical statistics were reorganized.

From her experience in the Crimea, Nightingale believed that medical treatments should be compared statistically to find out which were the most effective. Also, hospitals should be compared to determine which were better or worse, so that poor ones could be improved. She saw that she needed statistical data for this purpose, but found that the various English hospitals had different standards of record keeping, and that they classified and named diseases differently, so that data from their records could not be combined. She recognized that statistics are not useful unless they have a common base and are reliable. To encourage the gathering of usable data, she devised a uniform mode of reporting to be used by hospitals. She included a list of defined diseases, drawn up with Farr's advice and based on the classification used at the General Registry Office.[22]

An International Statistical Congress met in London in 1860. At the Congress a paper by Nightingale was presented outlining her scheme of uniform hospital statistics; copies of the form were distributed and discussed. Her plan was so well received that it gained her a nomination and election to the London Statistical Society (now the Royal Statistical Society); she was the first woman to be so honored.

Nightingale's form asked for the number of patients in a hospital at the beginning and end of the year, the number admitted during the year, the numbers discharged or deceased, the mean length of hospital stays, and other such data. Unfortunately the form she designed was lengthy and complex, and the classification of diseases, which was based on the inadequate knowledge of the time, was not widely accepted. As a result, although her general plan was a good one, the form was not generally adopted.

After the Indian mutiny of 1857, government inquiry produced reports of the appalling sanitary condition of the British army in India. Within a short time, Nightingale was pushing for an Indian Sanitary Commission. Her work in this field continued for the rest of her active life and led her into broad areas of public health, and beyond that to famine relief, land finance, education, and other fields.

Nightingale succeeded in getting a royal commission of inquiry to look into the state of health of the army in India. She designed a questionnaire to be sent to all the military stations. Her analysis of the returned forms resulted in a report to the commission of over 2,000 pages of small print. She found that the death rate of British soldiers in India was 69 per 1,000, three times as high as the mortality of troops in England before the sanitary reforms.[23] The death rate was six times higher than the rate among civilians in England. The high mortality resulted, not from the tropical climate, but from the same unsanitary conditions that had caused disease in the Crimea—overcrowding, bad drainage, contaminated water. Nightingale urged reform of sewage systems, barracks, and hospitals. The commission reported in 1863, with recommendations for change. Most of the report was her work.[24] The recommendations were put into effect, and 10 years later mortality had fallen from 69 to 18 per 1,000 men.[25]

Nightingale wanted sanitary reforms extended to the civilian population of India. She pushed for this with each successive Viceroy of India, educating them about the need for public health reform.

During the Statistical Congress of 1860 Nightingale met Adolphe Quetelet, who soon became her intellectual hero and guide. She was too ill to attend the meeting, but she received visitors, Quetelet among them.

Nightingale revered Quetelet from the time of their first

meeting. In 1872 she received from him the gift of a copy of the 1869 edition of his *Physique sociale*, which she read over and over, making marginal notes on every page. In his discovery of the statistical regularities of crimes, suicides, and marriages, she found confirmation of her belief that statistics revealed God's laws. Obviously, Quetelet had made a study of only certain limited aspects of human behavior, but for Florence Nightingale his work represented the beginning of the understanding of the relationship of God to man. She saw the book as a religious work— "a revelation of the Will of God."[26]

Ever since the Scientific Revolution of the seventeenth century, as we have seen, thinkers had hoped to find laws of human society that would in the social area be the analogs of the great laws of physical science such as the law of universal gravity. But by the middle of the nineteenth century it was obvious that such simple kinds of laws could not be found. Nightingale saw in Quetelet's work that the quest for sociological analogs of physical laws had been a failure and that the new science of "physique sociale" was statistically based.

It must not be thought, however, that in her great admiration for Quetelet, she never disagreed with him. An example of her critical thinking appears in her annotations to his work *Physique sociale*: she twice referred to a statement of Quetelet's as "nonsense." At issue was Quetelet's discussion of free will and the problem of whether the discovery of statistical necessity implied that free will does not exist. Quetelet's solution to this problem was to assume that free will does exist, but only within narrow limits, and has little effect when a large number of individuals is considered. For Nightingale, what mattered was the free will of administrators, who can choose whether to try to change the conditions that affect the way people act.

Nightingale thought *Physique sociale* a work of genius. At the same time she perceived how poorly organized it was, produced by his method of collecting separate essays, written at different times, without integrating them. She repeatedly urged Quetelet to produce a revised and better organized edition.[27] However, Quetelet (who had suffered a stroke years before and had never completely recovered from it) never undertook this. His death in 1874 saddened her. She and Farr raised money to support a memorial to Quetelet.[28]

A PASSION FOR STATISTICS

Nightingale's "passion for statistics," as she wrote to Quetelet, was "not in the least based on a love of science, a love I would not pretend I possessed." She had no pretension to being a scientist. Rather, her enthusiasm for statistics had two aspects that can only be understood in terms of her own personality and outlook.

First, she believed that statistics could be used to improve the condition of humanity. She had seen much of "the irrelevance of laws and of Governments, of the stupidity, dare I say it?—of our political system, of the dark blindness of those who involve themselves in guiding our body social,"[29] and only statistical studies, she thought, could guide the government aright.

For instance, statistics could be used to assess the effect of legislation. In his letters to the Grand Duke, Quetelet had emphasized this practical value of statistics: "A government in modifying its laws," he wrote,

> should collect with care documents necessary to prove, at a future date, whether the results obtained have answered their expectations. . . . When a penal law is modified, the influence of the change should be felt: if this influence leaves no trace, and the results of previous years continue to recur, the modification is without effect, and consequently illusory. When, on the contrary, the effects produced are marked, they teach

whether the modification has been advantageous or injurious.[30]

Nightingale cared about people and their welfare, and she believed that the government could only advance their welfare if its actions were based on statistical information. She was therefore concerned with social analysis, to enable the government to draft legislation or to repeal it if it proved ineffective.[31]

Of even more importance was Nightingale's belief that statistics revealed the relationship of God to man, and the character of God, if not His essence. As a mystic, she was convinced that the "essential character of God" is "a Universal Being who is Law."[32] The laws of God, those of the physical world and those governing man, could be found out, by experience or by research, including statistical research. She believed that it was the duty of mankind to do so, so that man could act in accordance with the divine plan and help to bring mankind to perfection.

The first of Nightingale's statistical reports was her "Notes on matters affecting the health, efficiency, and hospital administration of the British Army, founded chiefly on the experience of the late war." This was a volume of more than 800 pages, of large format.

One of the novel features of this report was the inclusion of statistical diagrams. Nightingale did not invent the statistical diagram, but she was an early pioneer in their use and a real innovator in their design. She wanted readers to get her statistical message, and she knew that tables of numbers are not very interesting and would probably not even be read by the average reader.

As Nightingale, while preparing her "Notes" on the health of the army for the royal commission, wrote to Sidney Herbert, the commission's chairman, "None but scientific men ever look

into the appendices of a Report. And this is for the vulgar public.
. . . Now, who is the vulgar public who is to have it? 1. The
Queen. . . ."[33] So she enlivened her presentation with diagrams.
Similarly, she commissioned illustrations of soldiers in their
barracks, and of Indian water carriers, for her "Observations on
the Sanitary State of the Army in India," commenting that the
Queen might look at her copy "because it has pictures."
The Queen was typical of a wider public in Nightingale's mind.

Quetelet took a similar view: he wrote that the study of statistics "is singularly facilitated by diagrams. A simple line allows
us to appreciate at a glance a succession of numbers which the
most subtle mind would find it difficult to retain and compare."
Diagrams, he wrote, "afford relief to the mind."[34]

Today, hardly a day goes by in which the local newspaper
does not have statistical diagrams to illustrate trends or situations
that would be difficult to show clearly by words or numbers
alone. The use of statistical diagrams is so widespread that it is
difficult to imagine that it was only in the middle of the nineteenth century that they came into use.

Nightingale and Farr did not agree on the inclusion of diagrams and illustrations in statistical reports. "We do not want
impressions, we want facts," he wrote to her in 1861, sounding
exactly like Mr. Gradgrind in Charles Dickens's bitter attack on
statistics, *Hard Times.* "You complain that your report would be
dry. The dryer the better. Statistics should be the dryest of all
reading."[35]

Nightingale's innovative diagrams were a kind of pie-chart—
circular with displays in the shape of wedge cuts (see figure 9.1).
They were printed in several colors. In her "Diagram of the
Causes of Mortality in the Army in the East," blue areas (measured from the center) represent deaths from "Preventible or Mit-
igable Zymotic Diseases" (essentially infectious diseases), red
areas represent deaths from wounds, black wedges represent

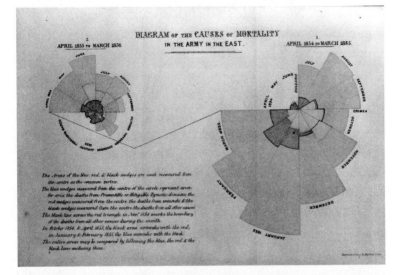

FIGURE 9.1 Nightingale's diagrams of the causes of mortality of the British army in the Crimean War, from her memorandum "Mortality of the British Army." *By permission of the Houghton Library, Harvard University*

deaths from all other causes. One sees at a glance that for most of the time the deaths from infectious diseases far outnumbered those from wounds or other causes. One can also see immediately how the mortality from each cause changed from month to month.

Diagrams of this kind are called Nightingale's "coxcombs," because of their shape and colors, which are somewhat like the brightly colored crest on the head of a cock. Nightingale herself used the term to refer to this report of the royal commission on the health of the army.[36]

So important did Nightingale consider these diagrams, which had originally formed an appendix to the report, that she reissued them as a separate publication. She also had a few copies of the diagrams framed and sent them to the War Office and other government offices, to keep the issues before their eyes.[37]

Florence Nightingale's belief that statistics revealed God's laws or plan for mankind led her to believe that people should be educated to know, understand, and appreciate statistical laws. She thought that administrators, especially, should have this knowledge. Administrators are trained at universities, and therefore universities should teach statistics. "We want to teach the men who are to govern the country the use of statistical facts."[38]

Nightingale wanted to establish a professorship of "applied statistics" at Oxford, in cooperation with Benjamin Jowett, Master of Balliol College. She tried to raise money for this purpose, and was prepared to give £2,000 of her own money to help endow the chair, as was Jowett. Francis Galton, who was brought into the plan, was less interested in teaching than in research, which Nightingale did not wish to support.[39] Ultimately the plan came to nothing.

In 1891, in a letter to Francis Galton, founder of eugenics, Nightingale wrote out a statement of the importance of statistics for the administrator or legislator. She instanced not merely sanitary reform, of the kind she had worked for, but questions of education, punishments for crime, efficacy of the workhouse system, and the health and economic welfare of the population of India. She concluded,

> You remember what Quetelet wrote [referring to his Letter 45 to the Grand Duke, quoted above] . . . "Put down what you expect from such and such legislation; after —— years, see where it has given you what you expected, and where it has failed. But you change your laws, your administering of them, so fast, and without inquiry after results past or present, that it is all experiment, see-saw, doctrinaire, a shuttlecock between two battledores."

In printing this letter in his biography of Galton, Karl Pearson wrote,

I confess—but then I am a prejudiced person, for the prophetess was proclaiming my own creed—that this letter appears to me one of the finest that Florence Nightingale ever wrote. What is more, it is almost as true to-day as it was thirty years ago. We are only just beginning to study social problems—medical, educational, commercial—by adequate statistical methods, and that study has at present done very little in influencing legislation. What is more, the requisite statistical teaching on which real knowledge must be based has hardly yet spread throughout our universities. The time has yet to come, when the want of a chair of statistical theory and practice in any great university will be considered as much an anomaly as the absence of a chair of mathematics. The logic of the former is as fundamental in all branches of scientific inquiry as the symbolic analysis of the latter.[40]

Florence Nightingale's passion for statistics is truly an indicator of the triumph of numbers.

EPILOGUE

Nightingale died 10 years after the twentieth century began, on the verge of a modern era that her innovations had helped to make possible. The changes the new century brought would be tremendous. The twentieth century not only opened up an impressive consolidation of the mathematical theory of statistics—above all, in probability theory—but it also totally transformed the amounts and kinds of numerical information available in virtually every area of human endeavor.

This previously unimaginable expansion of numerical information brought with it an increasingly urgent need for relieving the burden of repetitive arithmetical calculations. People had been looking for ways to make such calculations less burdensome since as early as the seventeenth century. In the eighteenth and early nineteenth centuries, a number of machines were invented and constructed that performed the four simple arithmetic operations. Of these the Difference Engine and the Analytic Engine of Charles Babbage (1791–1871) are the most interesting.

As we saw in the last chapter, a modified version of Babbage's Difference Engine was constructed and put to use in computing and printing out William Farr's *English Life Tables* (1864). The Analytic Engine, on the other hand, was even more sophisti-

Cover of 1890 *Scientific American* showing women census clerks handling the cards used by the Hollerith tabulating machines. *Courtesy of Widener Library of Harvard University*

cated, designed to have many features that we associate with today's computers. Sadly, no working model was ever actually built.

Yet, as the nineteenth century drew to a close, new means of mechanical calculation did make it possibly not only to calculate very large amounts of numerical information but also to tabulate impressive new *kinds* of data.

For example, the tabulation and correlation of the U.S. census of 1890—done 10 years before the death of Florence Nightingale—was performed by what were then and later known as "punch cards." (See figure on previous page.)

Herman Hollerith, a mining engineer, devised a system of punched cards to process the 1890 census. Holes punched at specific places on the cards represented the data. Electric contacts could meet through the holes, and by this means the holes could be counted.

Although the minimum data required for determining representation in Congress were not necessarily changed from earlier censuses, the census of 1890 generated and correlated social data of a kind never envisioned by the authors of the Constitution. The published report covers such diverse topics as school attendance, drains, cemeteries, paupers and the insane, cotton production, whaling boats and vessels owned in Boston, and the salaries and uniforms of the police of Austin, Texas.

Hollerith founded a company to exploit this technology, which later merged with two companies to form the present IBM (International Business Machines). Punched cards were later known as "IBM cards," and the history of the computer, as we now understand it, was well under way.

The rise of numbers had carried the world to the brink of a new "digital" age of information.

ACKNOWLEDGMENTS

Without the assistance of special colleagues, friends, and students as well as my wife, this book would not have been completed.

I owe a special debt to Sylvia Boyd, a lifelong friend who devoted hours of work in revision to get the book together, in many cases converting my rough notes into meaningful text.

Philip Davis Loring, my research assistant, helped by responding to specific research questions taking full advantage of the new computer technology. He solved research puzzles the keys of which were to be found on the shelves of Harvard's Widener Library. He was able to bring to a close many open questions raised by my work with Barbara Finan who as my research assistant worked closely with me for many months.

Elaine Storella, a professor at Framingham State College, ably assisted me. Her frequent visits to France helped us to solve many puzzles that involved books not easily available outside France.

Many friends and colleagues have come to share my enthusiasm for the different ways in which numbers have come to affect the conduct of government and of daily living as well the exact sciences. Chief among these is George Smith, Acting Director of the Dibner Institute. Peter Buck of Harvard University also generously offered to be a reader.

This subject was developed in the course of my teaching over many years and derived in its form under the important influence of my experience as a member of The Research Group under the direction of the late Lorentz Kreuger. Among the members of that group who were most influential in their thinking about probability and statistics were Rainie Daston, Ted Porter, and Stephen Stigler.

In retrospect a most important influence on the development of my ideas are the writings of Ian Hacking. His ideas gave direction to my thinking and show how a negative series of comments can challenge and change one's own formulations.

I am ever grateful to Judy Lajoie, administrator of the History of Science Department at Harvard University, for her help in providing tools of research and writing without which my jumble of research could never have been converted into this book.

Finally, and in many ways most importantly, I gladly acknowledge the most generous offer of my colleague and close friend, George Smith, to read the proofs for this book.

I did want to add in acknowledging the assistance of my co-workers that I am more than ordinarily grateful since in the last months my declining health and loss of vision have presented difficult challenges to the completion of this work.

NOTES

CHAPTER 1
1. Archibald (1949).
2. Edwards (1949), p. 117.
3. Ray (1998).
4. Lehner (1997).
5. Herold (1962); Aubry (1954).
6. This account of Dr. Parry's work is based on Parry (1997) and (1999).
7. Cobbett, *Parliamentary History of England, 1066–1803*, vol. 14 (1806), pp. 1330–31, quoted in Cohen (1982), pp. 231–50.
8. Cohen (1982), p. 35.
9. *New England Historical and Genealogical Register* 40 (1886), pp. 66–72, quoted in Cohen (1982), pp. 233–39.
10. 1870 History of the Census, *House Reports*, 41st Congress, 2nd session, quoted in Cohen (1982), p. 256, n. 24.
11. Besterman (1976), p. 168.

CHAPTER 2
1. Galileo (1914), p. 153.
2. Ibid., p. 179.
3. Kilgour (1954).
4. Leeuwenhoek to N. Grew, secretary of the Royal Society; reprinted in Leeuwenhoek (1948), pp. 3–35. This episode was brought to my attention by Cohen (1995).
5. Clark (1999), pp. 49–53.
6. Cook (1998), p. 116.
7. These scholars include the late William Coleman, Lorraine Daston, Ian Hacking, Theodore Porter, and Stephen Stigler. Their work does not render superfluous the older studies by Vincenz John, Isaac Todhunter, Helen Walker, and Harald Westergaard.

8. Brewer (1990).
9. Graunt (1662), p. 17.
10. Aubrey (1949), p. 238.
11. Kennedy (1983).
12. Graunt (1662), p. 24.
13. Ibid., p. 32; see also Westergaard (1968).
14. Graunt (1662), p. 68.
15. Greenwood (1948).
16. For more on Graunt and his contributions to statistics, see Keynes (1971) and Greenwood (1948).
17. Keynes (1971).
18. Ibid., p. vii.
19. Keynes (1971), p. 70.
20. Ibid.; see also Petty (1899), pp. lxxx–xci.
21. Pearson (1978).

CHAPTER 3

1. Gjertsen (1986), pp. 5–6, 550–551.
2. Westfall (1980), p. 36.
3. Schimmel (1993).
4. De Morgan ([1915], 1954), p. 55.
5. De Morgan ([1915], 1954), p. 373.
6. Ibid., p. 374.
7. Westergaard ([1932], 1968), pp. 32–33.

CHAPTER 4

1. U.S. Constitution, Article I, Section 2.
2. Wills (1978), p. 149.
3. Hutcheson (1725), p. 168ff. The equation shown here appears in the first edition of Hutcheson's *Inquiry*; it was deleted from later editions.
4. Ibid.
5. Hales (1727), p. xxvi.
6. Jefferson (1905), p. 194.
7. Jefferson (1944), p. 622ff.
8. See ibid., pp. 33–34 (entry for 31 July 1772).
9. Thomas Jefferson, Letter to James Madison, 30 January 1787, in Jefferson (1950–), vol. 11, p. 93.
10. Jefferson to James Madison, 6 September 1787, in Jefferson (1950–), vol. 15, p. 392.
11. Jefferson to James Madison, 20 December 1787, in Jefferson (1950–), vol. 12, p. 442.
12. U.S. Constitution, Article I, Section 2.
13. Balinski and Young (1982), p. 20.

14. Franklin to Peter Collinson [1752?] in Franklin (1959–), vol. 4, pp. 393–96.
15. This was the fourth English edition of Franklin's *New Experiments and Observations on Electricity* (London: D. Henry, 1754). The square, unfortunately, contained two typographical errors, both of which were fixed in the third French edition of 1773. For a detailed accounting of the various editions of the work, see Cohen (1941), pp. 139ff, and also Franklin (1959–), vol. 4, p. 393.
16. Franklin (1959–), vol. 4, p. 393. The reference is to a 1676 work by Bernard Frénicle de Bessy; see ibid., p. 393, n. 5.
17. Franklin (1959–), vol. 1, p. 150.
18. Ibid., p. 149.
19. Ibid., p. 149, n. 8.
20. Cassedy (1984).
21. Reprinted in Franklin (1959–), vol. 4, pp. 225–34.
22. Reprinted in Franklin (1959–), vol. 9, pp. 47–100.
23. "On the Death of His Son," originally printed in *The Pennsylvania Gazette*, 30 December 1736; reprinted in Franklin (1959–), vol. 2, p. 154.
24. Ibid.
25. Reprinted as "Preface to Dr. Heberden's Pamphlet on Inoculation," in Franklin (1959–), vol. 8, pp. 281–86.
26. Franklin to Catherine Ray, 16 October 1755, in Franklin (1959–), vol. 6, p. 225.

CHAPTER 5

1. Heilbron (1990).
2. Ibid.
3. Walker (1929), pp. 31–38.
4. Heilbron (1990), p.11.
5. Ibid.
6. Heilbron (1990).
7. Partington (1957).
8. Lavoisier's *Traité Elémentaire de Chimie* (Paris, 1789), p. 140, translated in Partington (1957), p. 124.
9. Westergaard (1932), pp. 97–98.
10. Perrot, Introduction to Lavoisier (1988), pp. 85, 88.
11. Lavoisier (1988), pp. 148–51.
12. Westergaard (1932), p. 98.
13. Ibid., p. 99.
14. Mitchison (1962).
15. Mitchison (1962), pp. 124, 134.
16. Jevons (1878).
17. Armitage (1983), p. 322.

18. Weiner (1993), p. 257.
19. Armitage (1983), pp. 321–34.
20. Ackerknect (1953), cited in Armitage (1983), p. 322.
21. Armitage (1983), p. 322.
22. Ibid.
23. Philippe Pinel, *Mémoires de l'Académie des Sciences* 8 (1807), p. 169, cited in Coleman (1982), pp. 131–32.
24. Westergaard (1968), p. 147.
25. Armitage (1983), p. 321.
26. Westergaard (1968), p. 174.
27. Quoted by Coleman (1982), p. 133.
28. Armitage (1983), p. 324.
29. Hacking (1990), p. 79.
30. Steiner (1940), p. 454.
31. Ibid., p. 453.
32. Cassedy (1984), p. 66.
33. Bartlett's *Essay,* cited by Cassedy (1984), p. 66.
34. Finan (2001), p. 55.
35. Shattuck papers, 24 October 1844, cited by Warner (1986), p. 298.
36. Osler (1908), p. 133.
37. Ibid., p. 132.
38. Coleman (1982), p. 133.
39. *Dictionary of Scientific Biography* on "Condorcet."

CHAPTER 6

1. Coleman (1982).
2. From a summary of Guerry's work in Bulwer (1834), p. 132.
3. Hacking (1990), p. 76.
4. Bulwer (1834), p. 148.
5. Funkhouser (1937), p. 304.
6. Breton (1987), pp. 191–92.

CHAPTER 7

1. Hacking (1990), p. 105.
2. Westergaard ([1932], 1968), p. 170.
3. Ibid., p. 166.
4. Landau and Lazarsfeld (1968), p. 248.
5. Ibid., p. 251.
6. Sarton (1935).
7. Porter (1985).
8. Wolowski, quoted in Hankins (1968), p. 29.
9. Landau and Lazarsfeld (1968), p. 250.
10. Quetelet (1984), p. 69.
11. Ibid.

12. Quetelet (1984).
13. Ibid., pp. 19–20.
14. Ibid., p. 19.
15. Quetelet ([1842], 1969), p. 118.
16. Hankins (1968); Lottin (1912).
17. Hacking (1990), p. 105.
18. Quetelet ([1849], 1981), Letter 20.
19. Landau and Lazarsfeld (1968), p. 249b.
20. Ibid., p. 250a.
21. Porter (1994).
22. Herschel (1857), pp. 384–85.

CHAPTER 8
1. Carlyle ([1839], 1974), pp. 124, 126.
2. Campbell and Garnett (1882), p. 646.

CHAPTER 9
1. Pickering (1974).
2. Cook (1913), vol. 1, p. 17.
3. Pickering (1974), p. 100.
4. Cook (1913), vol. 1, p. 130.
5. Verney (1970), Introduction.
6. Cook (1913), vol. 1, p. 134.
7. Cook (1913), vol. 1, pp. 146–47.
8. Nightingale ([1859] 1992), pp. 8, 18, 19 fn, 45, 53.
9. Eyler (1979), pp. 167, 174.
10. Cook (1913), vol. 1, p. 178.
11. Small (1998), cited in Epstein (2001).
12. Cook (1913), vol. 1, p. 314.
13. Eyler (1971), p. 267, as cited by Diamond and Stone (1981), p. 69.
14. Epstein (2001).
15. Ibid.
16. Merzbach (1977), p. 29.
17. From the inaugural issue of the *Journal of the London Statistical Society*, quoted in Eyler (1979), p. 15.
18. Details of Farr's career may be found in Eyler (1979). A summary is given in Diamond and Stone (1981), p. 69.
19. Diamond and Stone (1981), p. 68.
20. Farr to Nightingale, 14 February 1857, quoted in Diamond and Stone (1981), p. 69.
21. Cook (1913), vol. 1, p. 316.
22. Eyler (1979), p. 179.
23. Nightingale (1863).
24. Kopf (1916), p. 56.

25. Cook (1913), vol. 2, p. 182.
26. Cook (1913), vol. 1, p. 480.
27. Diamond and Stone (1981), part 1, p. 72.
28. Ibid., p. 74.
29. Ibid, p. 72.
30. Quetelet (1849), Letter 45.
31. Nightingale to Quetelet, 18 November 1872, cited in Diamond and Stone (1981), part 1, p. 72.
32. Cook (1913), vol. 1, p. 480.
33. Nightingale to Herbert, Christmas Day 1857, quoted in Diamond and Stone (1981), part 1, p. 70.
34. Quetelet (1849), Letter 39.
35. Unsigned note in Farr's handwriting, c. March 1861, cited in Diamond and Stone (1981), part 1, p. 70.
36. Small (1998).
37. Cook (1913), vol. 1, pp. 366–67.
38. Letter to Benjamin Jowett (1891), quoted in Cook (1913), vol. 2, p. 396.
39. Cook (1913), vol. 2, p. 397.
40. Pearson (1924), vol. 2, p. 418.

LITERATURE LIST

The following list includes the sources for quotations, events, and ideas. It is not intended to give a complete bibliography of the subjects of the book. In many cases, where textual material derives from reading in many sources, I limited the citations wherever possible to works in English and primarily those readily accessible.

ACKERKNECHT, ERWIN H. 1948 "Hygiene in France, 1815–1848." *Bulletin of the History of Medicine* 22, no. 2, pp. 117–55.

———. 1950. "Elisha Bartlett and the Philosophy of the Paris Clinical School." *Bulletin of the History of Medicine* 24, pp. 43–60.

———. 1952. "Villermé and Quetelet." *Bulletin of the History of Medicine* 26, pp. 317–29.

———. 1967. *Medicine at the Paris Hospital 1794–1848* (Baltimore: Johns Hopkins Press).

ARCHIBALD, RAYMOND CLARE. 1949. *Outline of the History of Mathematics* (Menasha, WI: Mathematical Association of America).

ARMITAGE, P. 1983. "Trials and Errors: The Emergence of Clinical Statistics." *Journal of the Royal Statistical Society, Series A (General)* 146, no. 4, pp. 321–34.

AUBREY, JOHN. 1949. *Brief Lives.* (London: Secker and Warburg).

AUBRY, PAUL V. 1954. *Monge: Le Savant Ami de Napoléon Bonaparte, 1764–1818* (Paris: Gauthier-Villars).

AUSTRIAN, GEOFFREY D. 1982. *Herman Hollerith: Forgotten Giant of Information Processing* (New York: Columbia University Press).

BALINSKI, MICHEL L., AND H. PEYTON YOUNG. 1982. *Fair Representation: Meeting the Ideal of One Man, One Vote* (New Haven: Yale University Press).

BENJAMIN, B. 1968. "Graunt, John." In *International Encyclopedia of the Social Sciences*, vol. 6, ed. David L. Sills (New York: Macmillan Company and Free Press), pp. 253–55.

BESTERMAN, THEODORE. 1976. *Voltaire* (Chicago: University of Chicago Press).

BRETON, PHILIPPE. 1987. *Histoire de l'informatique* (Paris: Editions la Decouverte).

BREWER, JOHN. 1990. *The Sinews of Power: War, Money and the English State, 1688–1783.* (Cambridge, MA: Harvard University Press).

BUCKLE, HENRY THOMAS. 1859. *History of Civilization in England*, 2 vols. (New York: D. Appleton and Company).

BULWER, HENRY LYTTON. 1834. *France: Social, Literary, Political* (New York: Harper & Brothers).

CAMPBELL, LEWIS, AND WILLIAM GARNETT. 1882. *The Life of James Clerk Maxwell* (London: Macmillan and Co.).

CAMPBELL-KELLY, MARTIN, AND WILLIAM ASPRAY. 1996. *Computer: A History of the Information Machine* (New York: Basic Books).

CARLYLE, THOMAS. [1839] 1974. "Chartism." Reprinted in *The Works of Thomas Carlyle*, vol. 29, ed. H. D. Traill (New York: AMS Press).

CASSEDY, JAMES H. 1984. *American Medicine and Statistical Thinking, 1800–1860* (Cambridge, MA: Harvard University Press).

CLARK, GEOFFREY. 1999. *Betting on Lives: The Culture of Life Insurance in England, 1695–1775* (New York: St. Martin's Press).

COHEN, I. BERNARD. 1941. *Benjamin Franklin's Experiments* (Cambridge, MA: Harvard University Press).

———. 1948. *Science, Servant of Man: A Layman's Primer for the Age of Science* (Boston: Little, Brown).

———. 1967. "Galileo, Newton, and the Divine Order of the Solar System." In *Galileo, Man of Science*, ed. Ernan McMullin (New York: Basic Books).

———. 1984. "Florence Nightingale," *Scientific American* 250, no. 3 (March), pp. 128–37.

———. 1985. *Revolution in Science* (Cambridge, MA: Harvard University Press).

———. 1987. "Scientific Revolutions, Revolutions in Science, and a Probabilistic Revolution 1800–1930." In *The Probabilistic Revolution*, vol. 1: *Ideas in History*, ed. Lorentz Krüger, Lorraine J. Daston, and Michael Heidelberger (Cambridge, MA: MIT Press).

———. 1990. *Benjamin Franklin's Science* (Cambridge, MA: Harvard University Press).

———. 1990. "G.D. Cassini and the Number of the Planets: An Example of Seventeenth-Century Astro-Numericological Patronage." In *Nature, Experiment, and the Sciences: Essays on Galileo and the History of Science*, ed. Trevor H. Levere and William R. Shea (Boston: Kluwer Academic Publishers).

———. 1994. *Interactions: Some Contacts between the Natural Sciences and the Social Sciences* (Cambridge, MA: MIT Press).

———. 1995. *Science and the Founding Fathers: Science in the Political Thought of Jefferson, Franklin, Adams, and Madison* (New York: W. W. Norton & Company).

COHEN, JOEL E. 1995. *How Many People Can the Earth Support?* (New York: W. W. Norton & Company).

COHEN, PATRICIA CLINE. 1982. *A Calculating People: The Spread of Numeracy in Early America* (Chicago: University of Chicago Press).

COLEMAN, WILLIAM. 1982. *Death Is a Social Disease: Public Health and Political Economy in Early Industrial France* (Madison: University of Wisconsin Press).

COOK, ALAN. 1998. *Edmond Halley: Charting the Heavens and the Seas* (Oxford: Clarendon Press).

COOK, E. T. 1913. *The Life of Florence Nightingale*, 2 vols. (London: Macmillan and Co.).

CULLEN, M. J. 1975. *The Statistical Movement in Early Victorian Britain: The Foundations of Empirical Social Research* (New York: Barnes & Noble).

DEANE, PHYLLIS. 1968. "Petty, William." In *International Encyclopedia of the Social Sciences*, vol. 12, ed. David L. Sills (New York: Macmillan Company and Free Press), pp. 66–68.

DE MORGAN, AUGUSTUS. 1915. *A Budget of Paradoxes*, 2nd ed. (Chicago: Open Court Publishing Company; reprint, New York: Dover Publications, 1954).

DESROSIÈRES, ALAIN. 1998. *The Politics of Large Numbers: A History of Statistical Reasoning*, trans. Camille Naish (Cambridge, MA: Harvard University Press).

DIAMOND, MARION, AND MERVYN STONE. 1981. "Nightingale on Quetelet," *Journal of the Royal Statistical Society, Series A (General)* 144, part 1, pp. 66–79.

———. 1981. "Nightingale on Quetelet II." *Journal of the Royal Statistical Society, Series A (General)* 144, part 2, pp. 176–213.

———. 1981. "Nightingale on Quetelet III: Essay in Memoriam." *Journal of the Royal Statistical Society, Series A (General)* 144, part 3, pp. 332–51.

DICKENS, CHARLES. 1837. "Full Report of the First Meeting of the Mudfog Association for the Advancement of Everything." *Bentley's Miscellany* (October); reprinted in *The Dent Uniform Edition of Dickens' Journalism: "Sketches by Boz" and Other Early Papers, 1833–1839*, ed. Michael Slater (London: J. M. Dent, 1994).

DOSSEY, BARBARA MONTGOMERY. 2000. *Florence Nightingale: Mystic, Visionary, Healer* (Springhouse, PA: Springhouse Corp.).

DUKE, THOMAS S. 1910. *Celebrated Criminal Cases of America* (San Francisco: James H. Barry Co.).

EDWARDS, I. E. S. 1949. *The Pyramids of Egypt* (Harmondsworth, Middlesex: Penguin Books).

EPSTEIN, HELEN. 2001. "The Mysterious Miss Nightingale." *New York Review of Books* 48, no. 4 (March 8), pp. 16–19.

EYLER, JOHN M. 1979. *Victorian Social Medicine: The Ideas* *and Methods of William Farr* (Baltimore: Johns Hopkins University Press).

FINAN, BARBARA. 2001. "A Statistical Survey of the Lowell 'Mill Girls' in 1845." (Master's thesis, Harvard University).

FRANKLIN, BENJAMIN. 1754. *New Experiments and Observations on Electricity*, 2nd ed. (London: D. Henry and R. Cave).

———. 1959–. *The Papers of Benjamin Franklin*, 36 vols. to date (New Haven: Yale University Press).

FUNKHOUSER, H. GRAY. 1937. "Historical Development of the Graphical Representation of Statistical Data," *Osiris* 3 , pp. 269–404.

GALILEI, GALILEO. 1914. *Dialogues Concerning Two New Sciences*, trans. Henry Crew and Alfonso de Salvio (New York: Macmillan Co.; reprint, New York: McGraw-Hill Book Company, 1963).

GJERTSEN, DEREK. 1986. *The Newton Handbook* (London: Routledge & Kegan Paul).

GRAUNT, JOHN. 1662. *Natural and Political Observations Mentioned in a Following Index, and Made upon the Bills of Mortality* (London: Tho. Roycroft; reprint, New York: Arno Press, 1975).

GREENWOOD, MAJOR. 1948. *Medical Statistics from Graunt to Farr* (Cambridge, UK: Cambridge University Press).

HACKING, IAN. 1990. *The Taming of Chance* (Cambridge, UK: Cambridge University Press).

HALES, STEPHEN. 1727. *Vegetable Staticks* (London: W. and J. Innys, and T. Woodward; reprint, London: Oldbourne, 1961).

HANKINS, FRANK H. 1908. *Adolphe Quetelet as Statistician* (New York: Columbia University Press; reprint, New York: AMS Press, 1968).

HANKINS, THOMAS L. 1985. *Science and the Enlightenment* (Cambridge, UK: Cambridge University Press).

HARVEY, WILLIAM. 1952. *The Circulation of the Blood: And Other Writings*, trans. Robert Willis (New York: E. P. Dutton).

HEILBRON, J. L. 1990. "Introductory Essay." In *The Quantifying Spirit in the 18th Century*, ed. Tore Frängsmyr, J. L. Heilbron, and Robin E. Rider (Berkeley: University of California Press).

HEROLD, J. CHRISTOPHER. 1962. *Bonaparte in Egypt* (London: Hamish Hamilton).

HERSCHEL, JOHN F. W. 1857. "Quetelet on Probabilities." *Essays from the Edinburgh and Quarterly Reviews* (London: Longman, Brown, Green, Longmans & Roberts; reprint, New York: Arno Press, 1981).

HUFF, DARRELL. 1954. *How to Lie with Statistics* (New York: W. W. Norton & Company).

HUTCHESON, FRANCIS. 1725. *An Inquiry into the Original of our Ideas of Beauty and Virtue, Treatise 2, Concerning Moral Good and Evil* (London: J. Darby).

JEFFERSON, THOMAS. 1905. *The Writings of Thomas Jefferson*, ed. Albert Ellery Bergh, vol. 19 (Washington, DC: The Thomas Jefferson Memorial Association).

———. 1944. *Thomas Jefferson's Garden Book, 1766–1824*, ed. Edwin Morris Betts (Philadelphia: American Philosophical Society).

———. 1950–. *The Papers of Thomas Jefferson*, ed. Julian P. Boyd, 29 vols. to date (Princeton: Princeton University Press).

JEVONS, W. STANLEY. 1878. "Remarks on the Statistical Use of the Arithmometer." *Journal of the Statistical Society of London* 41, no. 4 (December), pp. 597–601.

JOHN, VINCENZ. 1884. *Geschichte der Statistik* (Stuttgart: F. Enke).

JOYCE, JAMES. 1934. *Ulysses* (New York: Random House).

KENNEDY, GAVIN. 1983. *Invitation to Statistics* (Oxford, UK: Martin Robertson).

KEPLER, JOHANNES. 1997. *The Harmony of the World,* trans. E. J. Aiton, A. M. Duncan, and J. V. Field. (Philadelphia: American Philosophical Society).

KEYNES, GEOFFREY. 1971. *A Bibliography of Sir William Petty, F.R.S. and of Observations on the Bills of Mortality by John Graunt F.R.S.* (Oxford, UK: Clarendon Press).

KILGOUR, FREDERICK G. 1954. "William Harvey's Use of the Quantitative Method." *Yale Journal of Biology & Medicine* 26 (April), pp. 410–21.

KOPF, EDWIN W. 1916. "Florence Nightingale as Statistician."

Publications of the American Statistical Association 15, no. 116 (December), pp. 388–404.

KRÜGER, LORENTZ, LORRAINE J. DASTON, AND MICHAEL HEIDELBERGER, EDS. 1987. *The Probabilistic Revolution*, vol. 1: *Ideas in History* (Cambridge, MA: MIT Press).

KRÜGER, LORENZ, GERD GIGERENZER, AND MARY S. MORGAN, EDS. 1987. *The Probabilistic Revolution*, vol. 2: *Ideas in the Sciences* (Cambridge, MA: MIT Press).

LANDAU, DAVID, AND PAUL R. LAZARSFELD. 1968. "Quetelet, Adolphe." In *International Encyclopedia of the Social Sciences*, vol. 13, ed. David L. Sills (New York: Macmillan Company and Free Press), pp. 247–57.

LAVOISIER, ANTOINE-LAURENT. 1988. *De la Richesse Territoriale du Royaume de France*, ed. Jean-Claude Perrot (Paris: Editions du C.T.H.S.).

LAZARSFELD, PAUL F. 1961. "Notes on the History of Quantification in Sociology—Trends, Sources and Problems." In *Quantification: A History of the Meaning of Measurement in the Natural and Social Sciences*, ed. Harry Woolf (Indianapolis: Bobbs-Merrill Company).

LEEUWENHOEK, ANTONI VAN. 1948. *The Collected Letters of Antoni van Leeuwenhoek*, vol. 3 (Amsterdam: Swets & Zeitlinger).

LEHNER, MARK. 1997. *The Complete Pyramids* (London: Thames and Hudson).

LOTTIN, JOSEPH. 1912. *Quetelet: Statisticien et Sociologue* (Louvain: Institut Supérieur de Philosophie).

MACKENZIE, DONALD A. 1981. *Statistics in Britain 1865–1930: The Social Construction of Scientific Knowledge* (Edinburgh: Edinburgh University Press).

MAILLY, ÉD. 1875. *Essai sur la Vie et les Ouvrages de L.-A.-J. Quetelet* (Bruxelles: F. Hayez).

MCCARTNEY, SCOTT. 1999. *ENIAC: The Triumphs and Tragedies of the World's First Computer* (New York: Walker and Company).

MCDONALD, LYNN. 2000. "An Unscrupulous Liar? Florence Nightingale Revealed in Her Own Writings." *Times Literary Supplement* (December 8), pp. 14–15.

MERZBACH, UTA C. 1977. *Georg Scheutz and the First Printing Calculator* (Washington, DC: Smithsonian Institution Press).

MITCHISON, ROSALIND. 1962. *Agricultural Sir John: The Life of Sir John Sinclair of Ulbster, 1754–1835* (London: Geoffrey Bles).

MOIVRE, ABRAHAM DE. 1756. *The Doctrine of Chances: or, A Method of Calculating the Probabilities of Events in Play*, 3rd ed. (London: A. Millar; reprint, New York: Chelsea Pub. Co., 1967).

NIGHTINGALE, FLORENCE. 1859. *Notes on Nursing* (London: Harrison & Sons; reprint, Middlesex: Scutari Press, 1992).

———. 1863. "Observations by Miss Nightingale on the Evidence Contained in the Stational Returns." *Royal Commission on the Sanitary State of the Army in India*, vol. 1 (London: George Edward Eyre and William Spottiswoode).

———. 1990. *Ever Yours, Florence Nightingale: Selected Letters*, ed. Martha Vicinus and Bea Nergaard (Cambridge, MA: Harvard University Press).

OSLER, WILLIAM. 1908. *An Alabama Student and Other Biographical Essays* (New York: Oxford University Press).

PARRY, R.G.H. 1997. "Rolling Stones." *Australian Academy of Technological Sciences and Engineering*, no. 95 (January–February).

———. 1999. "Assembling Pyramid Proof." *Ground Engineering* 32, no. 1 (January).

PARTINGTON, J. R. 1957. *A Short History of Chemistry*, 3rd ed. (New York: St. Martin's Press).

PEARSON, KARL. 1924. *The Life, Letters and Labours of Francis Galton*, vol. 2: *Researches of Middle Life* (Cambridge, UK: Cambridge University Press).

———. 1978. *The History of Statistics in the 17th & 18th Centuries against the Changing Background of Intellectual, Scientific, and Religious Thought*, ed. E. S. Pearson (London: Charles Griffin & Co.).

PERUTZ, M. F. 1999. "The Top Designer." *The New York Review of Books* 46, no. 7 (April 22), pp. 51–54.

PETTY, WILLIAM. 1899. "Political Arithmetick." In *The Economic Writings of Sir William Petty, together with Observations upon the Bills of Mortality More Probably by Captain John Graunt,* ed. Charles Henry Hull (Cambridge, UK: Cambridge University Press; reprint, Fairfield, NJ: Augustus M. Kelley, 1986).

PICKERING, GEORGE. 1974. *Creative Malady: Illness in the Lives and Minds of Charles Darwin, Florence Nightingale, Mary Baker Eddy, Sigmund Freud, Marcel Proust, Elizabeth Barrett Browning* (New York: Oxford University Press).

PORTER, THEODORE M. 1985. "The Mathematics of Society: Variation and Error in Quetelet's Statistics." *The British Journal for the History of Science* 18 (March), pp. 51–69.

———. 1986. *The Rise of Statistical Thinking, 1820–1900* (Princeton: Princeton University Press).

———. 1994. "From Quetelet to Maxwell: Social Statistics and the Origins of Statistical Physics." In *The Natural Sciences and the Social Sciences: Some Critical and Historical Perspectives, Boston Studies in the History of Science*, vol. 150, ed. I. Bernard Cohen (Boston: Kluwer Academic Publishers).

———. 1995. *Trust in Numbers: The Pursuit of Objectivity in Science and Public Life* (Princeton: Princeton University Press).

QUETELET, ADOLPHE. 1828. *Instructions Populaires sur le Calcul des Probabilités* (Bruxelles: H. Tarlier & M. Hayez; reprint, Roma: Istituto Nazionale di Statistica, 1996).

———. 1842. *A Treatise on Man and the Development of His Faculties* (Edinburgh: William and Robert Chambers; reprint, Gainesville, FL: Scholars' Facsimiles & Reprints, 1969).

———. 1849. *Letters Addressed to H.R.H. the Grand Duke of Saxe Coburg and Gotha, on the Theory of Probability,* trans. Olinthus Gregory Downes (London: Charles & Edwin Layton; reprint, New York: Arno Press, 1981).

———. 1984. *Research on the Propensity for Crime at Different Ages.* Trans. Sawyer F. Sylvester (Cincinnati, OH: Anderson Publishing Co.).

RAY, JOHN. 1998. "Pyramidiots at Play." *Times Literary Supplement* (13 March), p. 13.

ROSEN, GEORGE. 1955. "Problems in the Application of Statistical Analysis to Questions of Health: 1700–1880." *Bulletin of the History of Medicine* 29, pp. 27–45.

ROSENBERG, CHARLES E. 1979. "Florence Nightingale on Contagion: The Hospital as Moral Universe." In *Healing and*

History: Essays for George Rosen (New York: Science History Publications).

SARTON, GEORGE. 1935. "Preface to Volume XXIII of Isis (Quetelet)." *Isis* 23, no. 65 (June).

SCHIMMEL, ANNEMARIE. 1993. *The Mystery of Numbers* (New York: Oxford University Press).

SETTLE, THOMAS B. 1967. "Galileo's Use of Experiment as a Tool of Investigation." In *Galileo: Man of Science*, ed. Ernan McMullin (New York: Basic Books, Inc.).

SILLS, DAVID L. 1987. *Paul R. Lazarsfeld: A Biographical Memoir* (Washington, DC: National Academy Press).

SINCLAIR, SIR JOHN, ED. 1799. *The Statistical Account of Scotland, 1791–1799* (Edinburgh: W. Creech; reprint, East Ardsley, England: E. P. Publishing, 1977).

SMALL, HUGH. 1998. *Florence Nightingale: Avenging Angel* (London: Constable).

SMITH, F. B. 1982. *Florence Nightingale: Reputation and Power* (New York: St. Martin's Press).

STEINER, WALTER R. 1940. "Dr. Pierre-Charles-Alexander Louis, a Distinguished Teacher of American Medical Students." *Annals of Medical History* 2, no. 6 (November), pp. 451–60.

STIGLER, STEPHEN M. 1986. *The History of Statistics: The Measurement of Uncertainty before 1900* (Cambridge, MA: Harvard University Press).

STRACHEY, LYTTON. 1918. *Eminent Victorians* (London: Chatto and Windus).

SUTTON, SILVIA B. 1986. *Crossroads in Psychiatry: A History of the McLean Hospital* (Washington, DC: American Psychiatric Press).

THOMAS, KEITH. 1987. "Numeracy in Early Modern England (The Prothero Lecture)." In *Transactions of the Royal Historical Society* 37, pp. 103–32.

TODHUNTER, ISAAC. 1865. *A History of the Mathematical Theory of Probability from the Time of Pascal to that of Laplace* (Cambridge, UK: Macmillan and Co.).

TRUESDELL, LEON E. 1965. *The Development of Punch Card*

Tabulation in the Bureau of the Census 1890–1940 (Washington, DC: Bureau of the Census, Department of Commerce).

VERNEY, HARRY. 1970. *Florence Nightingale at Harley Street* (London: J. M. Dent & Sons).

WALKER, HELEN M. 1929. *Studies in the History of Statistical Method, with Special Reference to Certain Educational Problems* (Baltimore: Williams & Wilkins Company; reprint, New York: Arno Press, 1975).

WARNER, JOHN HARLEY. 1986. *The Therapeutic Perspective: Medical Practice, Knowledge, and Identity in America 1820–1885* (Cambridge, MA: Harvard University Press).

WEINER, DORA B. 1993. *The Citizen-Patient in Revolutionary and Imperial Paris* (Baltimore: Johns Hopkins University Press).

WESTERGAARD, HARALD LUDVIG. 1932. *Contributions to the History of Statistics* (London: P. S. King; reprint, New York: Agathon Press, 1968).

WESTFALL, RICHARD S. 1980. *Never at Rest: A Biography of Isaac Newton* (Cambridge, UK: Cambridge University Press).

WILLS, GARRY. 1978. *Inventing America: Jefferson's Declaration of Independence* (Garden City, NY: Doubleday).

WOODHAM-SMITH, CECIL. 1950. *Florence Nightingale, 1820–1910* (London: Constable).

INDEX

Index